Seán Thomas Barry
Chemistry of Atomic Layer Deposition

Also of Interest

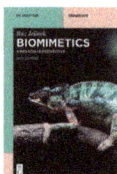

Biomimetics.
A Molecular Perspective
Raz Jelinek, 2021
ISBN 978-3-11-070944-5, e-ISBN (PDF) 978-3-11-070949-0,
e-ISBN (EPUB) 978-3-11-070994-0

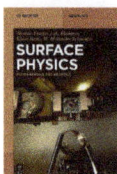

Surface Physics.
Fundamentals and Methods
Thomas Fauster, Lutz Hammer, Klaus Heinz, M. Alexander Schneider,
2020
ISBN 978-3-11-063668-0, e-ISBN (PDF) 978-3-11-063669-7,
e-ISBN (EPUB) 978-3-11-063699-4

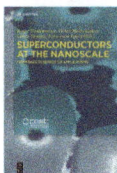

Superconductors at the Nanoscale.
From Basic Research to Applications
Roger Wördenweber, Victor Moshchalkov, Simon Bending, Francesco
Tafuri (Eds.), 2017
ISBN 978-3-11-045620-2, e-ISBN (PDF) 978-3-11-045680-6,
e-ISBN (EPUB) 978-3-11-045624-0

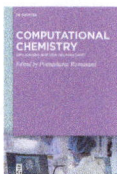

Computational Chemistry.
Applications and New Technologies
Ponnadurai Ramasami (Ed.), 2021
ISBN 978-3-11-068200-7, e-ISBN (PDF) 978-3-11-068204-5,
e-ISBN (EPUB) 978-3-11-068219-9

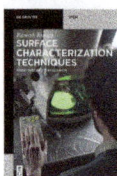

Surface Characterization Techniques.
From Theory to Research
Rawesh Kumar, 2021
ISBN 978-3-11-065599-5, e-ISBN (PDF) 978-3-11-065648-0,
e-ISBN (EPUB) 978-3-11-065658-9

Materials Science
Volume 1: Structure
Gengxiang Hu, Xun Cai, Yonghua Rong, 2021
ISBN 978-3-11-049512-6, e-ISBN (PDF) 978-3-11-049534-8,
e-ISBN (EPUB) 978-3-11-049272-9

Seán Thomas Barry

Chemistry of Atomic Layer Deposition

—

DE GRUYTER

Author
Prof. Seán Thomas Barry
Carleton University
Department of Chemistry
1125 Colonel By Drive
Ottawa K1S 5B6
Canada
sean_barry@carleton.ca

ISBN 978-3-11-071251-3
e-ISBN (PDF) 978-3-11-071253-7
e-ISBN (EPUB) 978-3-11-071259-9
DOI https://doi.org/10.1515/9783110712537

Library of Congress Control Number: 2021944243

Bibliographic information published by the Deutsche Nationalbibliothek
The Deutsche Nationalbibliothek lists this publication in the Deutsche Nationalbibliografie;
detailed bibliographic data are available on the Internet at http://dnb.dnb.de.

© 2022 Walter de Gruyter GmbH, Berlin/Boston
Cover image: farakos/iStock/Getty Images Plus
Typesetting: Integra Software Services Pvt. Ltd.
Printing and binding: CPI books GmbH, Leck

www.degruyter.com

This book is dedicated to my dear Antje. She knows why.

Foreword

Cherished Reader:

This textbook is not meant to be a comprehensive guide to the chemistry that can influence atomic layer deposition; that is, I think, unachievable. It also would not be instructive: mechanistic chemistry and chemical understanding does not come from remembering a specific reaction and recognizing it in a reaction, but rather by understanding the more fundamental aspects of chemistry and realizing how they apply. If you are like me, this process happens slowly.

For over two decades, I have taught, researched, read, and thought about atomic layer deposition and how chemistry applies to it. It has informed what I teach, and even what is important to me in my two principal fields of training: material science and chemistry. I have taught advanced undergraduate as well as graduate chemistry course in nanoscience and inorganic chemistry, and this textbook borrows the concepts from these classes that I consider important to understand atomic layer deposition. The first half of the textbook covers fundamental concepts that rely on organometallic chemistry, thermodynamics, and kinetics, while the second half of the textbook applies these fundamentals to specific deposition processes to promote a fuller understanding of the chemistry of atomic layer deposition.

I wanted to write a book that would have appealed to me as a new post-doctoral researcher: fundamental enough to give me a better insight into atomic layer deposition but appealing to my background education and experience in a way that would maybe spark an insight and creativity to allow me to innovate past what is shown in these pages.

I hope that this textbook can do that for you.

Cheers,
Seán

https://doi.org/10.1515/9783110712537-202

Contents

Chapter 1
Introduction

This textbook is written to be a wide-ranging course in the chemistry of atomic layer deposition (ALD). I envision this text to be used for courses in early graduate school, to introduce enough chemistry to new researchers that it helps clarify ALD, without overburdening them with too much unfamiliar knowledge. It comprises two main sections: the fundamentals of the ALD process, and selected reaction mechanisms from ALD, molecular layer deposition (MLD), and atomic layer etching (ALE).

The fundamentals of ALD include the essential chemistry of saturation and thermolysis, as well as the nature of ligands and precursors. These sections require a solid background in chemistry, and the underlying chemical concepts can be found in any university-level textbooks on physical chemistry and inorganic chemistry. The necessary background ranges from understanding structure and bonding, thermodynamics, and kinetics.

The sections on reaction mechanisms are based on examples to highlight particularly important or broadly applicable reactivities. These sections by their nature are not comprehensive and are not meant to be a review of the ALD process literature. Understanding several mechanisms can give a researcher insight when trying to understand new chemistry in their own ALD process. In my experience, proposing and studying mechanisms is a creative endeavor that mostly relies on postulating incorrect mechanistic pathways. Chemical reactivity occurs along many widely varied pathways, and a reaction mechanism is an attempt to make a linear narrative of these events. Ideally the sections on reaction mechanisms will provide enough of a foundation to spur better understanding of ALD processes.

1.1 Atomic layer deposition

ALD is a thin film deposition technique that has found its place in microelectronics manufacturing as well as a myriad of other applications. This technique is inherently a nanoscale technique since it allowed the fabrication of thin films developing – in a bottom-up fashion – from a molecular monolayer at a surface to a film of a target material. It is famous as a layer-by-layer, self-limiting, thin film growth method that relies on monolayer saturation of a surface through adsorption of a

Notes: In this chapter, I would like to acknowledge the work done by Prof. Riikka Puurunen, both in bringing to light the historical contributions of Russia to ALD, and for continuing to keep the fundamental descriptions of ALD phenomenon discussed by the community. Both her blog and "History of ALD" web pages are invaluable resources.

https://doi.org/10.1515/9783110712537-001

chemical precursor. The definition of the vocabulary of ALD is important: it is help-ful to the field in general if the practitioners of ALD all understand the necessary terminology:

1.2 Precursor

In general, a precursor is any volatile chemical compound that participates in the surface chemistry of an ALD process. Sometimes "reactant" is used, which is proper from a chemistry point of view, but this can also mean any chemical participating in any reaction, and so is too general. "Co-reactant" or "co-precursor" is less appro-priate: this implies that the chemical compound is added simultaneously with other precursors; the prefix "co-" should be avoided except when multiple, different chemical compounds are added at the same time.

1.3 Adsorption

This general term covers both chemisorption (which implies a lack of reversibility) and physisorption (which implies easy reversibility). I prefer it as a general term until a surface mechanism is understood sufficiently to be more specific. Chemi-sorption is often accompanied by a change in bonding in the precursor such that it cannot easily reverse along the same reaction pathway, perhaps through loss of a ligand as a volatile by-product. Physisorption often describes a low-energy interac-tion between a surface and a molecule, as a first mechanistic step that could lead to chemisorption.

1.4 Surface vs. substrate

Substrate (in ALD) is a borrowed definition from microelectronics. In general, the substrate is the surface on which deposition is initiated. Many processes start on silicon, making this surface the partner in the adsorption reaction. However, as de-position progresses, the "surface" becomes different from the "substrate." Once the substrate is entirely coated by the target film, ALD proceeds on this new surface. In some cases, the chemistry can drastically change when the surface changes. It bears remembering that the precursors and surface together comprise a thermody-namic system. Changing the surface (from initial substrate, to growing target film) naturally changes the thermodynamics of deposition.

1.5 Monolayer

Most ALD processes are considered to form a monolayer. This might conjure an image of a surface that is completely filled with adsorbed precursor molecules, but it truly means that the surface will not accommodate the further adsorption, regardless of whether the monolayer is densely packed or not. In most cases, the size of a precursor, as well as the number of nucleation sites, roughness of the surface, and other confounding factors all affect how densely made a monolayer is.

1.6 Self-limiting

This is the crux of ALD: that the monolayer, once formed, will stop adsorption, and remain stable until it is further reacted. In a perfect system, this would be a dense and chemisorbed monolayer that was thermodynamically stable under the conditions in which it was formed. But truly, an ALD process can exhibit self-limiting growth if the monolayer is kinetically stable, meaning that it persists long enough for the next chemical reaction to occur.

1.7 Layer-by-layer growth

The chemisorbed monolayer comprises a "layer," and the reaction of it with a second precursor further represents a "layer." The meaning of layer in layer-by-layer growth in the context of ALD is generally taken to mean a full layer of the target film. But, in a great majority of ALD processes, the growth-per-cycle is lower than what might be considered the thickness of one layer of material. This indicates that several different cycles might contribute to the same layer of target film. Given the variety of ways in which "layer" can be used in ALD, it is always worth further explaining what about a layer is being discussed.

It is commonly stated that ALD is a type of chemical vapor deposition (CVD), and the similarities are obvious. Both techniques require valving, furnaces, substrates and substrate holders, vacuum pumps, and the like for deposition to occur. But ALD is fundamentally different than CVD in the control that is exhibited over the precursor chemistry. In CVD, the principal reactants can react in the gas phase, at the surface, or in the continuum between those two different thermodynamic states. This allows for complex deposition chemistry to help produce the desired target film. An ideal ALD process does not allow for gas-phase interaction of the precursors, and idealized atomic layer deposition considers the reactivity at the surface to be paramount in the formation of the target film. This fundamental difference sets ALD apart as its own method. It is quite common for one precursor to be useful both as a CVD precursor and as an ALD precursor, and the insights in either deposition technique can influence the understanding of both.

It is in the chemistry of the precursor that the differences between ALD and CVD are highlighted. In CVD, more than one precursor is commonly mixed in the gas phase. Chemical reactions can occur in the gas phase and are commonly of benefit to the deposition of the intended target film. For example, boron carbide (B_xC) films can be deposited by CVD using triethylboron(III) as a single source precursor. However, the deposition temperature of this CVD process can be lowered to between 600–800 °C when dihydrogen (H_2) is used as a carrier gas and co-precursor. This is due to the reaction of triethylboron(III) with dihydrogen in the gas phase to produce highly reactive BH_3:

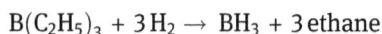

$$B(C_2H_5)_3 + 3H_2 \rightarrow BH_3 + 3\,\text{ethane}$$

It is the subsequent reactivity of the ethane that provides carbon for the B_xC films; this type of reactivity is precluded in ALD. In ALD, the precursors are necessarily introduced separately, where each can react to fully saturate a surface. When all precursors are available to the growth surface continually, the growth is also continual (i.e., CVD-type growth): there is not saturation of a monolayer of one precursor.

In the above case, the triethylboron(III) can act as a single-source precursor: all of the elemental components required for the deposition of the target film (B, C) are available in the precursor. The thermal reaction of the triethylboron(III) produces two precursors in situ, which then allows the continual deposition of B_xC as the target film. Naturally, single-source precursors are not viable in ALD. Self-limiting growth requires each independent precursor to undergo chemistry with the surface to further the deposition of the target film.

ALD relies entirely on serial chemical reactions for growth to occur, and therefore can be carried out in a robust fashion with simple equipment. At its heart, ALD is a cyclical process where two or more precursors are sequentially entrained into a reactor, separated by steps that eliminate the by-products of reaction, and other reactants from the gas phase (Figure 1.1). This allows each precursor to interact with the surface independently and achieve a chemical reaction that is (ideally) unhindered by the complication of a variety of chemical pathways to continue. This separation of species allows control over the surface chemistry that drives ALD. This is why ALD can be used to deposit such a wonderfully diverse number of materials.

The classic idea of an ALD cycle bears some scrutiny: in Figure 1.1a, both "n" and "m" are used as variables to demonstrate that the surface reactions do not have to proceed through a loss of a single group (as is generally shown, like in Figure 1.1b). It also demonstrates that, in a two-precursor process, that the precursors must have a stoichiometric balance to eliminate all ligands from both precursors. Further chapters show specific ALD cycles, and it would be a good exercise to try to compare them to Figure 1.1a and consider what the values of "n" and "m" are in real, applied

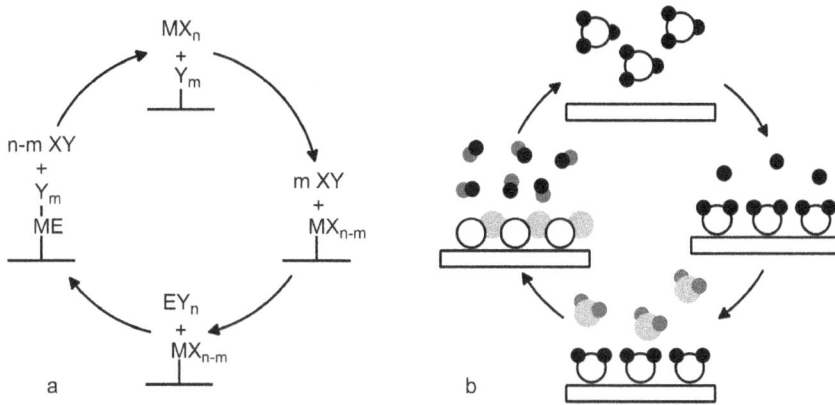

Figure 1.1: The classic depictions of an ALD cycle, where a) shows a stoichiometric balanced cycle, and b) shows a cartoon depiction.

ALD processes. In Figure 1.1b, the cartoon depiction of ALD is much more familiar, and (while maintaining a stoichiometric balance of "circles") it more easily depicts the concepts of chemisorption, by-product formation, and the separation of gas-phase vs. surface compounds. Both depictions are valuable in developing a holistic view of ALD.

1.8 Atomic layer deposition history

Historically, ALD as a technology originated from Finland in 1974 when the first patent application was submitted (then called "atomic layer epitaxy"). However, the scientific underpinnings of this technology go farther back and can be attributed to independent development by Tuomo Suntola in Finland, and earlier by Valentin Borisovich Aelskovskii in Russia.

Work done in Russia in the 1960s by the group of Aelskovskii under the name "molecular layering" was first published in the Russian literature in 1963, although it builds on Aelskovskii's 1952 concept of "the matrix" (presented in his doctoral thesis) where a framework of solid material would bear a stoichiometric number of surface nucleation sites in relation to its "bulk" (i.e., not surface) atoms. These nucleation sites could undergo chemistry to generate a chemically altered surface. This naturally evolved into the concept that, when an appropriate chemical was used to react with the nucleation sites in the framework model, another surface would be generated, again generating a stoichiometric number of nucleation sites. Already, the fundamental, cyclical nature of ALD can be seen to emerge in this research. Molecular layering was first described by Aelskovskii in a full paper in 1965 in *J. Appl. Chem. USSR*, and much of what might be considered early chemical ALD conceptualization was reported in the 1971 doctoral dissertation by Stanislav Ivano-vich Kol'tsov, a student of Aelskovskii. It was also in 1971 that Kol'tsov and Aelskov-skii submitted a patent application about molecular layering, entitled "the Principles of Synthesis of Solids," which was declined. It is interesting to study the development of molecular layering in Russia because it was framed primarily as a chemical ques-tion about altering the surface of a material through chemical means to produce dif-ferent kinds of surfaces and different ratios of bulk atoms to nucleation sites.

Concurrent with the molecular layering patent application and thesis in Russia, researchers in Finland were undertaking surface reactions to study the deposition of thin films. Although superficially this appears the same as molecular layering, this technology was engineered to be an industrially relevant solution to the fabri-cation of devices and was highly focused on the physics and engineering of thin film deposition and characterization. This research led to a successful patent appli-cation "Method for Producing Compound Thin Films," filed in 1974 by Tuomo Sun-tola and Jorma Antson. Under the name "atomic layer epitaxy," Suntola and Antson developed the technical foundation for ALD including concepts for selective area ALD and spatial ALD. It is interesting to note that the original patent contained de-tailed technical drawings of a reactor, highlighting their focus on physics and engi-neering. It is also interesting that Suntola defended this patent application in Moscow in the late 1970s, and the patent was granted in the Russian jurisdiction. The relationship to molecular layering as a prior art was not presented by the patent examiners, and an opportunity was missed to conjoin these two concepts at this ear-lier stage.

I find it reasonable that the Russian and Finnish developments occurred with-out collaboration or even knowledge of one another. The Russian molecular layer-ing was focused on the chemistry of a surface compared to the bulk, while the Finnish atomic layer epitaxy was focused on the engineering of thin film technolo-gies. It was not until 2004 that an effort was started to bring these two fields of study together to better understand the beginnings of ALD as we know it today. There has been a tremendous amount of work done on understanding the history of ALD, and Riikka Puurunen can be credited with bringing the early Russian con-tributions to the forefront.

Atomic layer deposition became the name of this method in the 1980s, and it was initially focused on chalcogenide thin films (i.e., containing a group 16 element like sulfur or selenium). This was due to the early influence of Suntola, who studied these types of films for electroluminescent displays. The field expanded into oxides and ni-trides in the 1990s, and during this period, research efforts expanded from a core of researchers based at Finnish institutions to the rest of the world. The ALD community became larger and more interconnected through the 1990s, and likewise the topics of interest in ALD evolved tremendously from chalcogenide electroluminescent displays and oxide and nitride MOSFET transistor technology – which remains important – into other fields like photovoltaics, solar cells, batteries and storage, memory, and catalysis. The versatility of the technique to make a variety of materials can be attributed to this success.

A perfect ALD process would involve thermodynamically stable monolayers of chemisorbed precursors reacting with other, selected chemicals delivered from the gas phase. Ideally, complications of gas-phase reaction and reversible chemisorption are permissible in ALD. This, of course, is too rigorous and confining of an ideal. ALD monolayers are rarely, if ever, thermodynamically stable, and even the most naïve understanding of chemistry permits reversibility in all reaction states. For this textbook, ALD will be defined as a technique where the stability of a monolayer is *sufficient* to allow the introduction of a different precursor for reaction in the timescale of the ALD cycle.

An ALD precursor is often idealized as the first compound introduced in an ALD cycle, or as the compound that contains "the important" component of the target film. For example (and an example that will be continually revisited), the common precursor for alumina (Al_2O_3) deposition is trimethylaluminum(III) (TMA). This is accurate, in that TMA delivers the "payload atom": alumina cannot be deposited without a source of aluminum. However, TMA is used in many different variations of alumina deposition, and the second precursor can be water, ozone, peroxide, oxygen plasma, or the combination of these. In all these examples, the oxygen-containing discrete chemical species is also a precursor. In this textbook, "precursor" will be used for each independent chemical species necessary for the deposition of the intended target film.

The surface is also a component of the necessary chemical reactions that drive an ALD process. The chemistry of atomic layer deposition requires a precursor to undergo reaction with two distinct surfaces over the lifetime of a deposition process. Initially it must react with surface nucleation sites to form a monolayer of adsorbed molecules that, upon reaction with subsequent precursors, initiates the deposited film. As that film coalesces, subsequently reaction occurs at nucleation sites on the growing film (hearkening back to Aelskovskii's "matrix" reactivity) which necessarily has a different surface chemistry than the initial substrate. There are several examples where a growing surface can cause chemical reactivity that the substrate does not. For example, many gold compounds will catalytically decompose on gold metal but not necessarily on other metals or binary extended solids. Many potential gold precursors can nucleate quite well on an oxide substrate, but once gold metal forms, uncontrolled surface reactions dominate and lead to CVD growth. This growth no longer self-limits. Likewise, transition metal oxide surfaces can catalytically decompose metalorganic precursors, preventing deposition by poisoning the surface with decomposed precursor fragments.

Primarily, the reactivity of precursors at a substrate surface requires a nucleation point. Chemically, this is most commonly a defect at the surface bearing a proton that can undergo Bronsted acid chemistry, with the proton being abstracted by a basic part of the incoming precursor to eliminate a conjugate acid (in the form of a protonated ligand), subsequently causing the gas-phase precursor to chemisorb at the surface in a relatively irreversible fashion. This newly formed surface chemical species now becomes part of a growing monolayer that will cover the surface. Once the monolayer is completely formed (i.e., when all accessible nucleation sites are taken up by chemisorbed species from the gas phase), then the surface is ready for reaction with a second precursor, which ideally forms a monolayer of the target film. What is interesting here and what sets it apart from CVD is that the only chemical reactions that are necessary for film deposition are between surface groups and precursors entrained from the gas. Although it is possible that the gas-phase precursor might undergo a spontaneous chemical reaction, ALD is ultimately driven by the thermodynamics of the surface reaction.

1.9 Trimethyl aluminum and water

In later chapters, this text will go through the specifics of ligand and precursor design, and what makes a good precursor. As an introduction, the very-well-studied trimethylaluminum(III) and water will be shown in depth here.

TMA is a very simple organometallic compound to make. In a synthetic laboratory equipped to handle air-sensitive materials, this precursor can be made on the gram scale by reacting trichloroaluminum(III) and methyllithium(I), both of which can be purchased from chemical suppliers:

$$\text{ALCL}_{3\,(\text{sol})} + 3\,\text{MeLi}_{(\text{sol})} \rightarrow \text{“ALMe}_3\text{”}_{(\text{sol})} + 3\,\text{LiCL}_{(\text{s})}$$

There are three aspects of the reaction as written that bear further explanation. First, this reaction must be done in a non-protic solvent. This means that the solvent cannot have a hydrogen atom that is accessible as an acidic proton. In this case, both reagents would preferentially react via Bronsted acid/base chemistry with the acid. Generally, methyllithium(I) is sold in a solvent like diethylether (a polar, but non-protic solvent); this would be a good solvent in which to carry out the synthesis of TMA. Second, the reaction should be carried out in a solvent that does not dissolve the co-product lithium chloride. If this salt can be isolated as a solid and precipitate out of the solution where the TMA is being made, this will cause the reaction to go all the way to completion (this is an example of "Le Chatelier's principle"). This type of reaction (that is thermodynamically driven to completion by the formation of a salt) is called a "salt metathesis" reaction, and is one of the most common reactions employed in synthetic inorganic chemistry.

For large-scale and industrial uses, TMA is made from metallic aluminum and methyl chloride. The methyl chloride reacts with the aluminum to make a variety of methylchloroaluminum(III) species, and sodium is then used to activate the methyl chloride to complete the reaction to TMA. The overall reaction produces TMA and NaCl (table salt):

$$2\,Al + 6\,CH_3Cl + 6\,Na \rightarrow Al_2(CH_3)_6 + 6\,NaCl$$

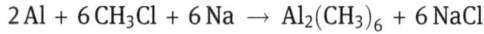

The third point about the synthesis of TMA also concerns the use of it as a gas-phase precursor. Because the aluminum wants to have 8 valence electrons, it will not form as a discrete "AlMe$_3$" unit, but rather it will double up to form the "dimer" hexamethyldialuminum(III) (Figure 1.2).

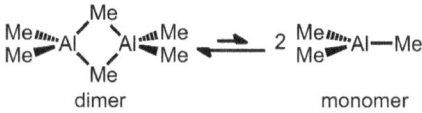

dimer   monomer

Figure 1.2: The dimer-monomer equilibrium of trimethylaluminum(III) lies to the dimer side of the reaction at room temperature, and at typical ALD delivery temperatures.

This dimer is in equilibrium with two equivalents of the monomer TMA, and this equilibrium is affected by the medium surrounding the molecules. For instance, at 100 °C in the gas phase, there is 80% dimer and 20% monomer when TMA is delivered at low pressure. Although ALD is commonly thought to occur with no gas-phase reactions, this simple equilibrium demonstrates this to not necessarily the case: hexamethyldialuminum(III) undergoes dissociation to produce TMA, which is the chemical species that easily chemisorbs at a surface nucleation site. Here the dissociation of TMA into monomers is necessary to permit reaction of TMA with a surface hydroxyl (Figure 1.3).

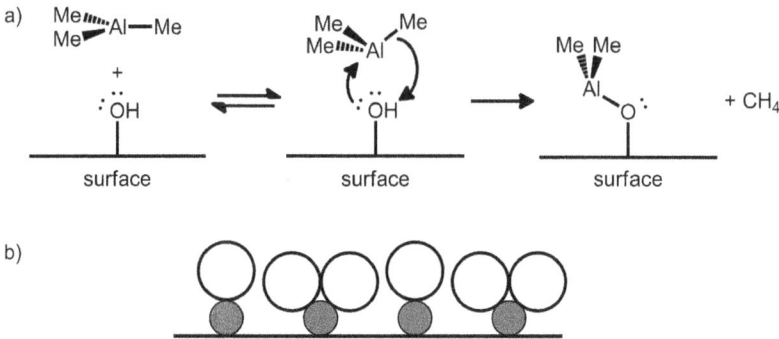

Figure 1.3: a) The reaction of TMA at a surface hydroxyl to release methane and chemisorb at the surface. b) The steric crowding of the methyl groups (white circles) preventing close packing of the aluminum centers (grey circles).

In order to fully understand the surface adsorption mechanism, we have to imagine every minute step of bonds being made and broken. As a monomer, TMA is a Lewis acid (i.e., will accept electrons), and so it will first physisorb to a hydroxylated surface through interaction with a lone pair of electrons from the surface oxygen. The reaction then proceeds by the methyl group making a bond to the proton on the hydroxyl and eliminating methane, causing chemisorption of a fragment of the TMA molecule. Although the reaction is shown losing one methyl, the surface reaction can occur between multiple hydroxyls. In general, TMA will chemisorb through two hydroxyls on a bare silicon oxide surface, and as an aluminum oxide surface grows, the chemisorption of TMA tends to happen only through one hydroxyl.

Reacting at the surface to lose one or more methyl groups makes the TMA fragment difficult to remove from the surface. This strong chemical bond holds the fragment in place, and since the methane can easily migrate away in the gas phase, the reaction becomes non-reversible (i.e., it cannot reclaim the lost methyl group). At the surface, these TMA fragments takes up more room than a lone aluminum atom would. This is generally used to describe why the growth per cycle (GPC) of aluminum oxide is around 1.1 Å, where a monolayer of aluminum oxide is around 2.9 Å. This is typical for ALD, the GPC of the process is generally less than the thickness of one "layer" of target film. It is helpful to remember this for ALD processes: due to either the density of nucleation sites at a surface, or the steric crowding of the ligands, a saturated monolayer does not represent a monolayer of the target film. The GPC of aluminum oxide from TMA and water is 30–40% of the target aluminum oxide film, depending on factors like temperature and nucleation density. It is also important to remember that the surface of growing aluminum oxide is some combination of TMA molecules that have lost either one or two methyl groups.

The surface reactivity that produces alumina from TMA and water is more complex than depicted here. Indeed, simple models of ALD surface reactivity might help idealize how film growth proceeds but can often disguise the complexity of the surface reactivity. Something as straight-forward as the depiction of layer-by-layer growth can serve as an example of how the complexity of the surface chemical model of ALD has evolved from its original conception. Looking back at Kol'tsov's 1971 thesis, a depiction of surface chemistry was included, and it resembles a lattice of atom points (Figure 1.4a). More modern depictions of ALD monolayer surfaces comes from Riikka Puurunen in 2020, and considers the size of the chemisorbed precursor, and the complexity of the surface reactions (Figure 1.4b). It might not be necessary in every cartoon of ALD to depict the complexities that can occur at the surface, but it helps remind a reader that there is more than just material science to consider when thinking about ALD.

a)

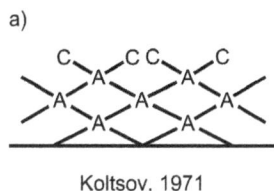

Koltsov, 1971

b)

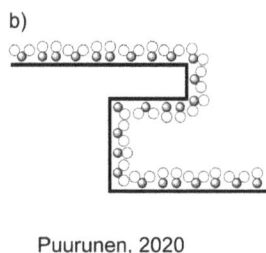

Puurunen, 2020

Figure 1.4: The evolution of the depiction of layer-by-layer growth, from a) "molecular layering" to b) "atomic layer deposition." This figure is adapted from the original figures.

1.10 Helpful resources

At the time of writing, there are several helpful online resources that can complement topics in this book. These are three principal blogs that cover the history of ALD, as well as up-to-date lists of processes, precursors, methods, and publications, and news and industrial announcements.

1.   The "Virtual Project on the History of ALD" (https://vph-ald.com): this website is curated by Riikka Puurunen, and has several editable documents concerning the early work in both molecular layering and atomic layer epitaxy. It contains reading lists for new researchers in the field, and comprehensive publications on the history of ALD. There is a related blog (http://aldhistory.blogspot.fi), also curated by Riikka Puurunen.

2.   "Atomic Limits" (https://www.atomiclimits.com/): this website is administered through the Plasma & Materials Processing group at Eindhoven University of Technology, led by Erwin Kessels. There are topical blog posts on many aspects of atomic layer processing in general, and a main contribution are the searchable databases for ALD and ALE processes. This resource is useful for every practitioner of ALD, and can provide the opportunity for a self-guided study of the field.

3.   "BALD Engineering" (http://www.blog.baldengineering.com): this blog is an up-to-date archive of industrially relevant atomic layer processing news and analysis. It is curated by Jonas Sunqvist and provides wide-ranging, and often entertaining, news about the field of ALD and beyond.

Chapter 2
Saturation

Saturation and saturative growth are the cornerstone of ALD; this is what differentiates the technique from other chemical deposition methods, and is also the underlying reason that ALD can be performed in relatively unsophisticated equipment: the nature of the chemical reaction of the precursor and the surface governs the deposition, not physical parameters of the deposition like flow and pressure, although these physical parameters have secondary effects on the surface chemistry, and are necessary technical issues from the point of view of macroscopic uniformity and employment of a process in a practical sense. But the surface chemistry and, importantly, the layer-by-layer growth mechanism rely heavily on the thermodynamics of surface reactions. When ALD is typically explained, it is described as depositing "a monolayer per cycle," which defines the growth per cycle (GPC). These characteristics can easily confuse and mislead a person new to the field and bear significant explanation.

The growth per cycle in ALD should be measured in units of depth (typically Angstroms or nanometers) and should not be referred to as the "growth rate." To do so would imply a timescale in the measurement, and a typical ALD cycle can vary wildly in the timing, with many timed sub-steps affecting the process. For instances, the simplest ALD cycle (an "AB" cycle) consists of two precursors introduced independently, separated by purge steps (Figure 2.1).

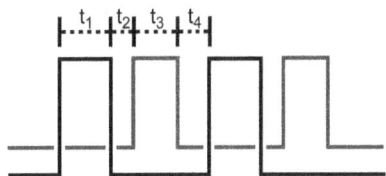

Figure 2.1: An idealized "AB" ALD cycle with four independent time variables.

If each square-wave line represents the pressure of a different precursor in the gas phase, then the whole cycle is seen as a sum of each of these pulse durations, plus the time it takes to clear each pulse from the reactor. It is common to optimize the pulse of the first precursor (t_1) in publications of ALD process development, and it is increasingly common to see optimization of the second precursor pulse (t_3) as well.

Notes: In this chapter, I would like to acknowledge the work done by Dr. Simon Elliott, who has often taken concepts in ALD and distilled them to their fundamental physical chemistry concepts. I have learned a lot from Simon, and several figures in this chapter are directly influenced by tutorials he has given.

https://doi.org/10.1515/9783110712537-002

It is less common to see optimization of the purge times (t_2 and t_4). This is an important step: a precursor like water – commonly used in oxide deposition – has a strong propensity to adsorb to the reactor interior, in places where deposition may not be occurring. This could be, for example, the stainless-steel chassis of the chamber. A long purge step using both evacuation by vacuum pump and purging with a heated gas flow might be necessary to remove all traces of water. The consequence of neglecting this optimization might be the inadvertent mixing of the independent precursors, by having the water stay in the chamber while the other precursor is pulsed. This would lead to CVD conditions: deposition occurring in the presence of both precursors and the possibility of gas-phase reactions. Water in this case can be contrasted with another common oxygen source, ozone (O_3), which both is far less likely to stick to reactor surfaces, and rapidly decomposes to O_2. A short purge step that is suitable for ozone would be unacceptable for water. For instance, during alumina deposition, a normal purge time for water in a typical showerhead reactor is on the order of 10 seconds while ozone can be eliminated in 1 second.

As well, the concept that a monolayer entirely forms during each cycle is an unsophisticated (and optimistic) view of growth, and this will be explored more in depth in this chapter. However, typical GPC values should be enough to demonstrate this to be incorrect. At typical deposition temperatures (200 °C – 250 °C) for alumina, the TMA/water process deposits approximately 1 Å per cycle compared to the thickness of a "monolayer" of alumina, which is approximately 2.9 Å. Note that it bears to be circumspect about the thickness of "one monolayer of alumina" here. Alumina is typically amorphous when deposited by ALD, and this necessarily comes with some ambiguity of structure. Even if the approximation of the alumina crystalline unit cell was used, the hexagonal (wurzitic) form of alumina has three Al-O planes per unit cell, with a height (in the z-axis direction) of 12.991 Å. It is very difficult to determine what the definition of one monolayer of alumina is given these complications. However, even an aluminum atom has a covalent radius of 1.18 Å, and so a "complete" alumina monolayer would have to be much thicker than 1 Å.

Another aspect to be concerned about when considering GPC is the fact that these values are often reported simply by measuring bulk thickness and dividing by the number of cycles used to achieve this thickness. This is a reasonable but idealized concept of GPC, and neglects considerations of surface chemistry, and changes in the surface that occur during deposition. A simple ALD process, through the time that it runs, will encounter two fundamentally different growth surfaces. Again, using the alumina as an example, the growth typically starts on a non-alumina surface. Indeed, most ALD process development will use silicon wafers among the sample surfaces for growth. These "coupons" (i.e., the small test wafer used) will have a 4 nm or thicker native oxide passivation layer of silicon oxide, unless they have been specially treated to otherwise not. Thus, many processes are developed on a silica surface. As the process continues, and alumina forms a continuous coating, the surface chemistry can differ drastically as growth must then occur on alumina.

In this example growth occurs on two archetypal surfaces: first silica, and then alumina. These two surfaces are quite different. Both surfaces are typically terminated in hydroxyl ("OH") groups where chemisorption is likely to occur; silica will be dominated by "tetrahedral" OH groups, while alumina can have both tetrahedrally supported and octahedrally supported hydroxyls (Figure 2.2). In the case of silica, there are commonly either one (mono-hydroxyl) or two (gem-hydroxyls) at a silicon atom. At 350 °C, there are approximately 3.5 hydroxyls per square nanometer, with a ratio of about 100:6 in favor of mono-hydroxylated Si atoms. This suggests that the reactivity at the surface will be driven predominantly by mono-hydroxyl nucleation sites, and lead to less complicated surface reactivity. For alumina, there five fundamental surface groups, divided between tetrahedrally coordinated (T_d) and octahedrally coordinated (O_h) groups (Figure 2.2). At 350 °C, there are between 5 and 7 hydroxyls per square nanometer, distributed among these geometries. Additionally, the ratio of these groups is dependent on the surface treatment to prepare the alumina surface. In all, the surface chemistry of alumina could lead to much more divergent chemical reactivity than on silica.

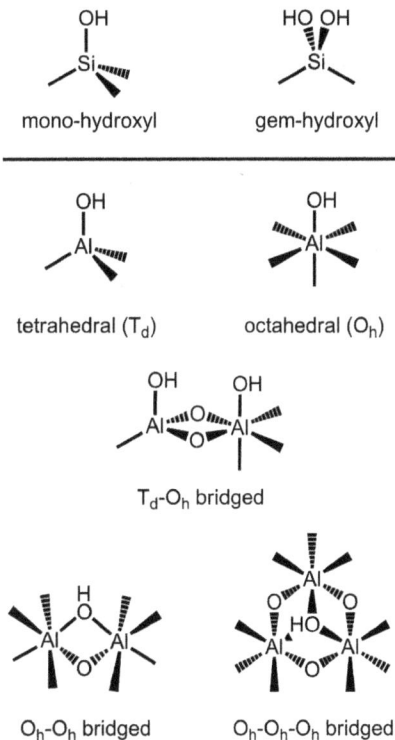

Figure 2.2: The various hydroxyl environments on either a silica or an alumina surface.

Complicating this idea of surface hydroxyl density is the changeable nature of surfaces, particularly with temperature. It is common for oxide surfaces to undergo water elimination and to produce bridging "oxo" groups (i.e., an oxygen) rather than a hydroxyl:

$$2\!\vdash\!OH \rightarrow (\vdash)_2O + H_2O$$

This dehydration becomes more prevalent with temperature, and silica can be completely denuded of hydroxyl groups at approximately 900 °C. Additionally, the presence of neighboring hydroxyl groups is known to increase the quantity of adsorbed water at a hydroxyl surface: close set hydroxyl groups can cooperate to strongly hydrogen bond water. As the hydroxyl density decreases through dehydration reactions, so does the amount of adsorbed water. Controlling the hydroxyl density at a surface by preheating can help tune nucleation and reactivity in the initially cycles of an ALD process.

Hydroxyl density needs to be considered in all processes: for instance, to determine how different is the surface chemistry in the initial (i.e., nucleation) cycles compared to the surface chemistry once the target film becomes the deposition surface. If the GPC differs drastically in the initial cycles of a process, the common method of calculating GPC (i.e., dividing final thickness by number of cycles) will have an error that becomes more egregious as the measured film becomes thinner. It is not uncommon to see a nucleation delay, where growth in the initial cycles is lower than during much of the deposition (Figure 2.3). Conversely, it is also possible to have nucleation enhancement, where deposition in the initial cycles is thicker than in most of the process. When many cycles are used to determine growth rate, errors caused by different GPC during the nucleation section will have less affect on the overall growth rate, but determining GPC for very thin films deposited by fewer cycles will amplify this error.

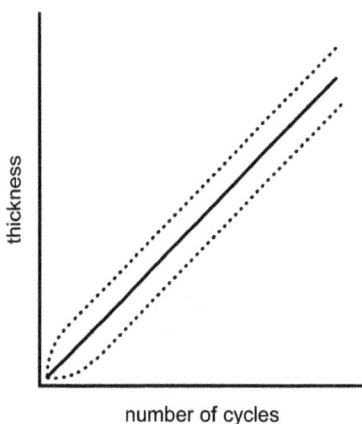

Figure 2.3: Nucleation delay or enhancement compared to uniform growth during all cycles.

Ideally, a process and its GPC can be measure in situ, and this can uncover nucleation issues and better inform the entire process. However, determination of GPC by simple division is so common that nucleation issues are initially missed and are brought to light only when a process is characterized in greater depth.

Saturation typically occurs through chemisorption of a precursor at a surface. A subsequent chapter will deal with the concept of saturation in depth, but here, the change in the precursor to its surface bonded state is important. When a molecule encounters a surface, it can undergo a chemical change. The most common is an exchange reaction. The precursor undergoes physisorption, bringing it into proximity of the surface. When the molecule forms a covalent bond at the surface to become chemisorbed, it will commonly lose a ligand to the surface, or into the gas phase. The remaining fragment of the precursor (hereafter I will refer to it as the "surface moiety") can have several ligands left intact, which can then provide the opportunity to continue the process to deposit the target film, but also cause impurities to be left in the film. However, a chemisorbed monolayer is not the end of surface organization. The monolayer itself undergo further organization to become more well-packed to facilitate higher coverage or could undergo a surface-mediated chemistry to eliminate surface moieties and cause lower coverage. It is reasonable, then, to discuss an ALD process in terms of kinetic stability rather than thermodynamic stability.

In the timeframe of an ALD cycle, the GPC can be show as a function of temperature (Figure 2.4). This commonly leads to the concept of the "ALD window," where the GPC is constant over a temperature range. This is a common feature of ALD processes, but it is not a necessary condition for the existence of an ALD process. The GPC represents the reaction of a stable monolayer, and given that the stability of the monolayer, and indeed, the formation of the monolayer itself is dependent on kinetics, the concept of an ALD window presupposes that there is necessarily a temperature range where the rate constant is independent of temperature for both monolayer formation (adsorption) and monolayer stability (desorption). This is not a necessary condition for ALD processes, and saturative growth can occur at different temperatures with different GPC. This can make diagnosing saturation by changing temperature difficult and demonstrates that a saturation curve is the only valid demonstration of self-limiting behavior.

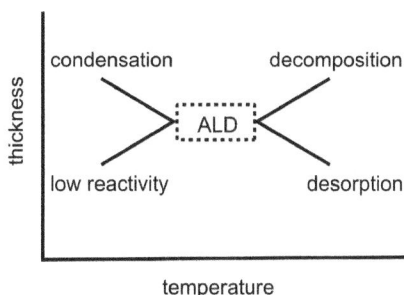

Figure 2.4: The variation of GPC with temperature.

At low temperatures, there are two potential non-saturating conditions: a low GPC that increases with temperature is typically indicative of the precursor-surface interaction not being given enough energy to allow saturation. A high GPC that decreases with temperature is indicative of the condensation of the precursor at the deposition surface. At high temperatures, a decreasing growth rate is indicative of either the desorption of the chemical monolayer or etching of the monolayer by the second precursor (or by a by-product of the surface reaction). An increasing growth rate is indicative of the thermal decomposition of the monolayer, which is CVD-like growth.

2.1 The saturation curve

The saturation curve for ALD is typically depicted as GPC vs. time, where the time represents the pulse time of one of the precursors (Figure 2.5). This is a common shortcut since the x-axis in a saturation curve necessarily represents the *exposure* of the surface to the precursor. Exposure (sometimes called "dosage") is derived from the kinetic molecular theory of gases (KMT) and has the units of 10^{-6} Torr. In general, the unstated assumption of a saturation curve is that the vapor pressure of the precursor is remaining constant, while the time is altered. It is also quite common that the pressure of the precursor delivered to the process conditions is not directly measured, and thus units of time "stand in" for the measured exposure.

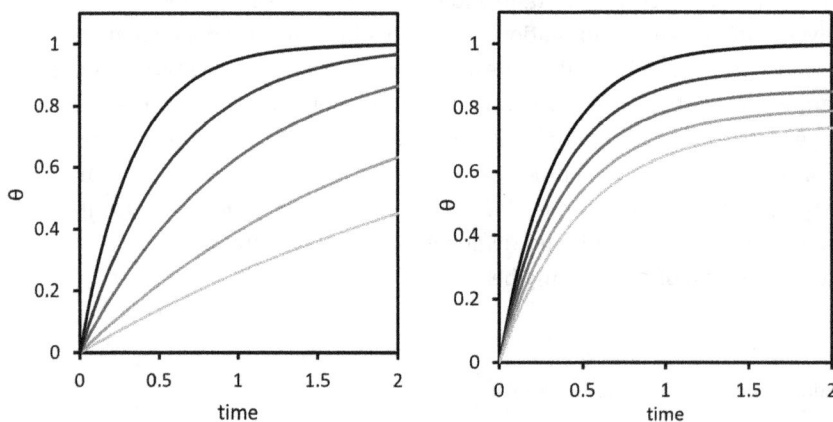

Figure 2.5: The effect of irreversible adsorption on saturation. a) Each successively darker line represents an increase in precursor pressure. b) Each successively darker line represents a lower k_d.

Saturation can occur when a species can be chemisorbed to a surface for a length of time greater than the time it takes to undertake a pulse sequence. The ideal saturating monolayer is (arguably) thermodynamically stable so that once it forms, it will remain unchanged until acted upon by a second precursor. But this is not a requirement for saturative growth. The monolayer needs to be stable only until the process can be continued (by reaction with another precursor, for example). The concept of the saturation curve derived below is based on kinetics, and its general formula can be derived from first principles.

This derivation starts by considering the adsorption and desorption of a gaseous precursor species "A" from a surface "⊢":

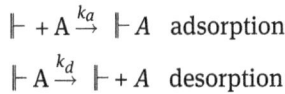

$$\vdash + A \xrightarrow{k_a} \vdash A \quad \text{adsorption}$$
$$\vdash A \xrightarrow{k_d} \vdash + A \quad \text{desorption}$$

Here, adsorption is a second-order elementary kinetic reaction (where the rate of absorption is R_a) that can be represented mathematically as:

$$R_a = k_a[A][\vdash]$$

and desorption is a first-order elementary kinetic reaction (where the rate of desorption is R_d):

$$R_d = k_d[\vdash A]$$

Here, the square brackets denote that we are discussing the concentration of the entity in question. To simplify this to use only one variable (and make it easier to handle mathematically), we can define the "chemisorption coverage" as θ. This will represent the fraction of the surface that is covered at any point:

$$\theta = [\vdash A]$$

Thus, the number of free sites available for a to occupy ("⊢") becomes:

$$1 - \theta = [\vdash]$$

so that the total of occupied and unoccupied surface sites is equal to one (i.e., 100%). The final change we need to make to the elementary kinetic reactions is to redefine the concentration of A in terms of pressure. This is not strictly necessary for the derivation but does frame the results in terms of recognizable process parameters. Here, we will assume that the gaseous precursor species "A" behaves like an ideal gas:

$$p_A = \frac{n_A RT}{V}$$

where "R" is the ideal gas constant and T is the system temperature. This is a common definition used in chemical kinetics, and, here, limits us to a constant temperature. Already, it is obvious that any saturation curve derived from here will be "isothermal" (i.e., representing one temperature). It is also notable that we do not have to be limited to an ideal assumption about gaseous "A" and could use any thermodynamic equation of state; we are just going to use the easiest one. Now the redefined elementary rate equations are:

$$R_a = k_a p_A (1 - \theta) \text{ absorption}$$

$$R_d = k_d \theta \text{ desorption}$$

It is not uncommon for the solution of the saturation curve to be considered as the "Langmuir Isotherm" equation, although this is incorrect. However, it is valuable to see how the Langmuir equation can be derived from these same assumptions. $\theta_{eq} = \frac{Kp_A}{Kp_A + 1}$ The Langmuir equation comes from the assumption that adsorption and desorption have reached an equilibrium, and that the "steady-state approximation" can be applied. In short, the steady-state approximation says that, at an infinite time, the forward and reverse reactions have reached a dynamic equilibrium such that species can still adsorb and desorb, but this does not change the value of θ. In this condition, the definition $\theta = \theta_{eq}$ is applied. In the case of equilibrium, the forward rate (R_a) and reverse rate (R_d) are equal:

$$k_a p_A (1 - \theta_{eq}) = k_d \theta_{eq}$$

$$k_a p_A - k_a p_A \theta_{eq} - k_d \theta_{eq} = 0$$

$$\theta_{eq}(k_a p_A + k_d) = k_a p_A$$

$$\theta_{eq} = \frac{k_a p_A}{k_a p_A + k_d}$$

This solution for the Langmuir equation is not the standard form, since the Langmuir equation is stated in terms of the equilibrium constant for the overall reaction:

$$A + \vdash \ \leftrightharpoons \ \vdash A$$

In the case here, we can use the concentrations defined in the elementary rate equations to define the equilibrium constant:

$$K = \frac{[\vdash A]}{[\vdash][A]}$$

If we rearrange the elementary rate equations from above and substitute them in (and recall that $R_a = R_d$ because of the steady state approximation), we can get the expression:

$$= \frac{R_d/k_d}{R_a/k_a} = \frac{k_a}{k_d}$$

So, if we multiple our derivation of the Langmuir isotherm by k_d/k_d, we can reach the more recognizable form of this equation:

$$\theta_{eq} = \frac{k_a p_A}{k_a p_A + k_d} \left(\frac{k_d}{k_d} \right) = \frac{(k_a/k_d)p_A}{(k_a/k_d)p_A + (k_d/k_d)} = \frac{K p_A}{K p_A + 1}$$

However, this is not the solution we are looking for (we will use it later in our derivation, however). This tells us how the equilibrium shifts with changing pressure of our gaseous precursor A, with time extended to infinity to ensure that equilibrium will be established.

Considering the conditions that might give us saturation in an ALD process, we should impose the condition that adsorption happens to a much greater extent than desorption, specifically $k_a \gg k_d$. In the extreme of this condition, as k_d approaches zero ($k_d \to 0$), the equilibrium constant approaches infinity ($K \to \infty$). Under this condition, $\theta_{eq} = 1$. Although this satisfies the idea that saturation occurs fully in a system dominated by adsorption at infinite time, this does not help demonstrate the time-dependent process of saturation.

What we want to know from these elementary rate definitions is how the coverage θ changes with time. Since adsorption makes θ larger, and desorption makes it smaller, we can define an integral of the surface coverage:

$$\frac{d\theta}{dt} = R_a - R_d$$

$$\frac{d\theta}{dt} = k_a p_A (1 - \theta) - k_d \theta = k_a p_A - \theta(k_a p_A + k_d)$$

$$\frac{d\theta}{k_a p_A - \theta(k_a p_A + k_d)} = dt$$

The limits of the integral will obviously be from no coverage ($\theta = 0$) to some coverage θ over a range of time:

$$\int_0^\theta \frac{d\theta}{k_a p_a - \theta(k_a p_a + k_d)} = \int_0^t dt$$

The right-hand side of the integral resolves to "t," and the left-hand side can be solved using a standard integral:

$$\int \frac{dx}{ax + b} = \frac{1}{a} \ln(ax + b)$$

where $a = -(k_a p_A + k_d)$ and $b = k_a p_A$. It is worth noting at this point that k_a and k_d are constant since rate constants are temperature dependent, and we have previously defined the temperature as constant. This gives us the solution:

$$\left[\frac{1}{-(k_a p_A + k_d)} * \ln\left[-(k_a p_A + k_d)\theta + k_a p_A\right]\right]_0^\theta = t$$

With a little manipulation, this solution can be rearranged to isolate the logarithm so that an exponential can easily be made:

$$\ln\left[\frac{-(k_a p_A + k_d)\theta + k_a p_A}{k_a p_A}\right] = -(k_a p_A + k_d)t$$

$$\frac{-(k_a p_A + k_d)\theta + k_a p_A}{k_a p_A} = e^{-(k_a p_A + k_d)t}$$

Here, we can borrow our previously developed definition of the equilibrium coverage (i.e., the Langmuir equation):

$$-\frac{k_a p_A + k_d}{k_a p_A}\theta + \frac{k_a p_A}{k_a p_A} = e^{-(k_a p_A + k_d)t}$$

$$\left(-\theta_{eq}\right)\theta = e^{-(k_a p_A + k_d)t} - 1$$

$$\theta = \frac{e^{-(k_a p_A + k_d)t} - 1}{-\theta_{eq}}$$

This can be simplified into its final form: the conditions of saturation we stated above can be applied, where $\theta_{eq} = 1$ when $k_a \gg k_d$. This is a specific type of surface saturation, where the chemisorbed species is more stable than the gaseous species, so there is a thermodynamic driving force to complete the monolayer. Under these conditions, we can assume the contribution of k_d to the exponential factor is negligible, and can be ignored for simplicity's sake:

$$\theta \approx 1 - e^{-k_a p_A t}$$

These final conditions do not need to be imposed, and for systems where there is an appreciable equilibrium between the chemisorbed state and the free gaseous precursor, the more complex equation can be considered.

In the case of an adsorption-favored saturation, it is immediately apparent that the rate constant of adsorption, the pressure of the precursor, and the time of exposure play similar roles, modifying the exponent in the same way. Of course, there is no method to have direct control over the rate constant of adsorption (k_a), but both the pressure of the precursor and the time of exposure can be controlled by process parameters, again demonstrating that exposure (measured in Langmuirs) is the best practical way to change the saturation curve. If we simply graph the above equation in coverage vs. time, the effect of exposure is apparent (Figure 2.5a).

This graph shows the effect in saturation over an order of magnitude change in exposure, with the lightest line being low exposure, and the darkest line being high exposure. Here, each line represents a different partial pressure of the precursor (p_A), and saturation is more quickly achieved for the highest pressure compared to the lowest pressure. Note well that the assumptions necessary for this derivation (that the gas behaves ideally, and that chemisorption is so favored that desorption does not affect the equation) limit it to being a fundamental exercise and cannot be used as a model for saturation. In existing processes, the non-ideal behavior of the thermodynamic system, non-uniformity, and precursor depletion due to surface up-take play significant roles in saturation, as do a myriad of other considerations (like the partial pressures or carrier gases, by-products, and localized thermal changes due to reactions at the surface and even in the gas phase).

Interestingly, this saturation does not need to be irreversible (Figure 2.5b). When the more complex form of this kinetic equation is used, we can see that vary-ing the reverse rate constant from zero (black line) to equal to the forward rate con-stant (lightest grey) shows that saturation slows down, but equilibrium coverage is still established. It should be noted here that this does not denote a completely satu-rated, full monolayer in each case, but a fraction of the equilibrium coverage θ_{eq}, and each successively higher reverse rate constant plateaus at a lower coverage.

2.2 Selectivity and saturation

The two graphs in Figure 2.5 have a major implication for ALD and the saturation of a surface from a kinetics point of view. The equilibrium can lie to the reactants side of the chemical equation, meaning that the reaction can form products, but thermo-dynamically, reactants are favored. So, saturation can occur if the time frame is long enough even if a full monolayer is not thermodynamically favorable. This can account for growth rates that are much lower than the expected "thickness of one monolayer of target material," which is commonly seen for growth of metal films. As well, this shows that a small change in the precursor's vapor pressure can have a very large influence on saturation, even in conditions where a full monolayer is not thermodynamically favorable. Thus, the vapor pressure of the precursor becomes an exceptionally important characteristic to control.

2.3 Steric bulk and saturation

The number of available surface sites for saturation can be a function of many fac-tors. At room temperature, silica has 4–5 hydroxyls per square nanometer, while alumina has 15. Thus, the growth rate of alumina might be expected to be much higher than silica. In general, however, both oxides tend to grow by 1 Å/cycle, and the saturation of trimethylaluminum(III) plateaus at 30–40% of a monolayer. This can partially be attributed to the steric bulk of the precursor: the amount of space it occupies at a surface (Figure 2.6).

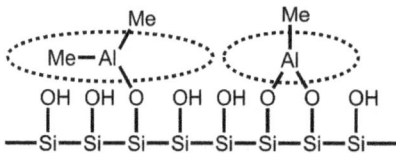

Figure 2.6: The steric influence of chemisorbed trimethylaluminum(III) fragments on a hydroxylated surface.

Remembering that the methyl groups in Figure 2.6 are three-dimensional, and are in constant rotational and vibrational motion, it is easy to imagine that each surface group exhibits an area of influence where it effectively "covers" nucleation sites and blocks them from reaction. It is also important to remember that chemisorption can follow several paths: in the above example, it is known that trimethylaluminum(III) nucleates at a silica surface through the loss of either one or two equivalents of methane, and so both resulting surface moieties would each exhibit a different area of influence. It is possible to model (computationally or geometrically) these areas of influence, and even simple calculations can help justify low growth rates caused by steric bulk.

2.4 Selectivity

Another important aspect that comes out of the analysis of the saturation curve using this derivation is the idea of selectivity. Specifically, here we can consider "kinetic selectivity." In the case of kinetic selectivity, two surfaces will show different saturation behavior due to their different reactivity toward the precursor. When different deposition surfaces exhibits two very different reactivities, we can assign each surface a different rate constant for adsorption. A common example of two different surfaces might be a hydroxyl-terminated oxide surface (common for silica and alumina for example), and a metal-terminated metallic surface (Figure 2.7). On such a surface, we will consider the hydroxyl-terminated section to react with a rate constant of k_a, and the metallic surface to react with k_a'.

In this case, we can consider the two reactions to occur on the same time axis with the same partial pressure of a precursor (p_A). Thus, the only difference in the fundamental definition of the saturation curves comparing the two surfaces is the difference in the rate constants that govern the kinetics of adsorption and desorption. When we consider the surfaces to differ in rate constant by two orders of magnitude (i.e., k_a:k_a' = 100:1), it becomes apparent that, at very short pulse times, the hydroxylated surface with the high rate constant (k_a) will show saturation while the metallic surface with the

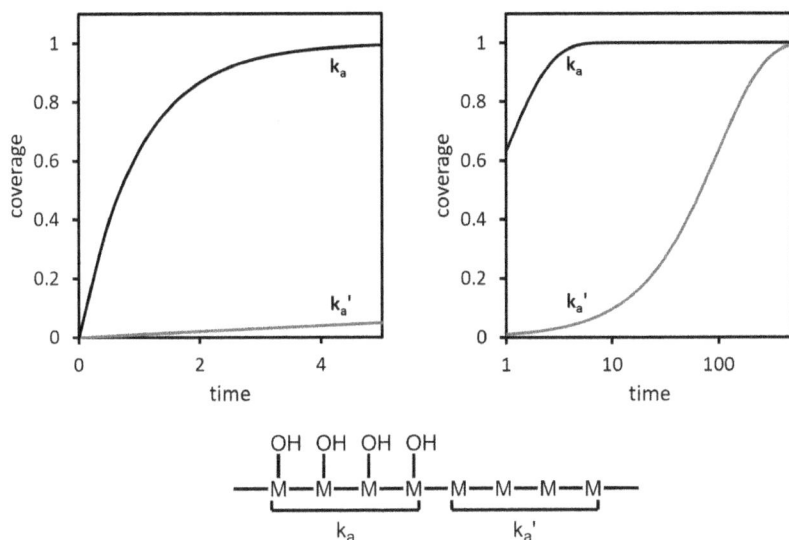

Figure 2.7: The effect of surface kinetics on deposition selectivity.

low rate constant (k_a') will show little or no growth. At longer much longer timescales (shown as a log scale for time in the bottom right graph of Figure 2.7), both surfaces could show saturation. Again, this is not a practical demonstration of selectivity, although it has been exploited in some processes, but a demonstration that kinetic selectivity is possible, where two surfaces where chemisorption is a spontaneous thermodynamic event can still be selective to deposition by the kinetics of precursor uptake.

There are many potential examples of saturation curves that could be used to illustrate the practical aspects of how saturation is achieved in a real experiment. To avoid (again) using alumina as an example, I will instead use an example from the research group of Prof. Henrik Pedersen in Linköping University, Sweden. This process was developed to grow indium nitride (InN) from trimethylindium(III) and ammonia plasma. InN is an important material in micro- and nano-electronics because it has a narrow bandgap, making it ideal for infrared lasers and other optical applications in electronics. In this example, the process can be seen to develop a growth plateau (i.e., an "ALD window") at 220 °C with a GPC of 0.4 Å, as well as at 300 °C with a GPC of 1.2 Å (Figure 2.8a).

When saturation was studied at these two temperatures, ALD growth was found in both cases, albeit with different GPCs (Figure 2.8b). This is an interesting case for saturation; there are very different levels of coverage occurring, but the experimental conditions ensure that the pressure of the incoming precursors (Me_3In and NH_3-plasma) is the same at all temperatures. Generally, both of these processes mostly grow on InN, which can be imagined as alternating between being In-terminated or

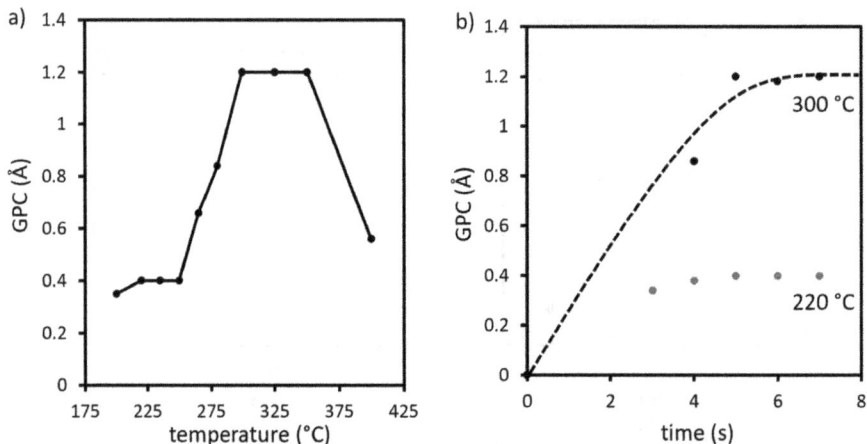

Figure 2.8: ALD study for InN. a) GPC vs. temperature and b) saturation of the same process at two different growths per cycle and temperatures.

N-terminated, depending whether Me_3In or NH_3 (plasma) was the last precursor to react at the surface. It could be that these two temperatures result in different surface densities of nucleation sites as discussed in Figure 2.2. However, it is reasonable to consider the number of nucleation sites to generally diminish as temperature increases. The difference between the adsorption rate constant (k_a) and the desorption rate constant (k_d) might have shifted significantly to prefer adsorption (as discussed in Figure 2.5). This is better reasoning since higher temperature could promote adsorption by permitting the reaction to perhaps overcome an activation energy at the surface. Contrarily, higher temperature could likewise promote desorption; an increase of temperature does not necessarily increase dose, and the large decrease in GPC from 350 °C and 400 °C is likely due to a rapid onset of desorption kinetics or thermal decomposition. These are conflicting potential reasons for the change in GPC, and this deposition process would need to be studied in greater depth in order to understand this interesting phenomenon. A final consideration is that this increase of temperature might be causing a thermal reaction in the precursors that is independent of the substrate (i.e., in the gas phase) and the increase of GPC is due to an unpredicted gas-phase reactant. This example illustrates the difficulty in determining the cause of ALD growth, even when it is identified with a saturation curve. Further experiments are often required to truly understand the nature of saturation and GPC.

Chapter 3
Ligands

In metal-organic chemistry, a ligand is an organic group that is bonded directly to a metal center. This bonding can be covalent or coordinative (described below), and it can be through one atom or many atoms. In general, precursor-ligand bonds need to be strong to withstand thermolysis, but need to be carefully considered, so that the chemistry of an ALD process can remove the ligands easily.

3.1 Ligand bonding

One of the main contributions that a chemist can make to ALD is to discover and implement a new ligand. A ligand is a molecule that can donate electrons to a central metal center, where the ligand acts as a Lewis base (i.e., electron donor), and the metal as a Lewis acid (i.e., electron acceptor) (Figure 3.1). This bonding to the metal is most easily envisioned as the overlap of the electron density of the ligand and the metal, and can be depicted as "orbitals," which are spatially distributed volumes where electron density is found. Although there are several levels of bonding theory that become increasingly complicated and accurate, herein we will consider "valence shell electron pair repulsion" (VSEPR) theory, with some more complicated aspects of bonding covered by molecular orbital (MO) theory. You can find these bonding models described in most elementary inorganic chemistry texts.

VSEPR theory results in "ball-and-stick" drawings, where the electron pair involved in bonding is represented by a line drawn between the two atoms. There is not very much nuance to this model, but it allows molecules to be depicted in a straightforward manner.

It is interesting to consider the VSEPR structure trimethylaluminum(III), one of the most fundamental precursors in ALD. Initially, we must consider what it means for a methyl group to be a ligand. Traditionally, a base (in acid-base theories) has a "conjugate" acid: the acid that occurs when a proton is added to the group. In the case of the methyl ligand, the conjugate acid is methane, an extremely stable molecule:

$$H_3C^- + H^+ \rightarrow CH_4$$

Note: In this chapter, I would like to acknowledge the undergraduate chemistry students at Carleton University: I developed courses in organometallic chemistry in collaboration with them, relying on their honest and curious critiquing of bonding concepts. It is true that you never really know a topic until you teach it, and that is thanks to the students.

https://doi.org/10.1515/9783110712537-003

a	b	c
L—M VSEPR	M—NH$_3$ **L** neutral ligand coordinative bond	M—NH$_2$ **X** anionic ligand covalent bond

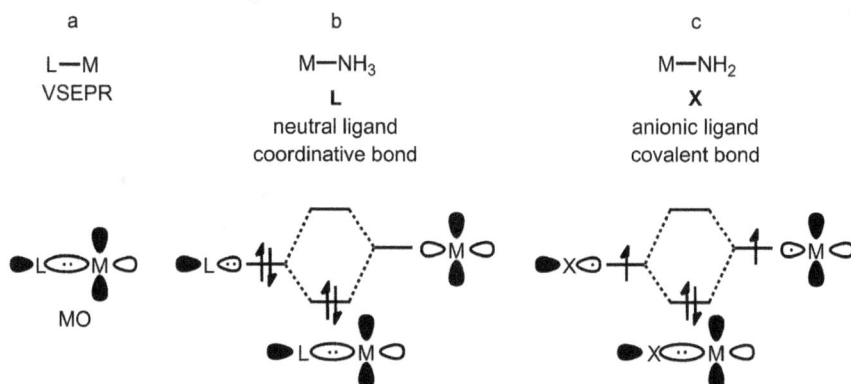

Figure 3.1: Ligand bonding. a) Two useful models to consider ligand-metal bonding. b) ligand "**L**" contributes both electrons to the M-L bond. c) ligand "**X**" and metal "**M**" each contribute one electron to the M-L bond.

The amount of energy needed to reverse this reaction, or to "deprotonate" a methane molecule to produce H$^+$ and CH$_3^-$, is quite high (440 kJ/mol), but it is not very different than the bond strength of the well-known strong acid, HCl (428 kJ/mol). It is not necessarily the bond strength that makes something acidic, but rather the electronegativity difference between the bonding pairs. Carbon (in the case of methyl) has an electronegativity of 2.55, which is very similar to that of hydrogen (2.20). Electronegativity is a scale of measurement of how much the atom "pulls" on the electrons in the bond. In a C-H bond, this tug-of-war is well matched. In the HCl bond, chlorine has an electronegativity of 3.16, and so can pull the electrons away from the hydrogen. So chlorine can more easily form "chloride" (the negatively charged ion of chlorine), and release H$^+$:

$$HCl \rightarrow H^+ + Cl^-$$

From the point of view of bonding to a metal center, a ligand is a Lewis base, meaning that it donates a pair of electrons (from a ligand orbital) to a metal orbital. There are two fundamental types of Lewis bases: those that have a pair of electrons to share as a neutral compound, and those that share a pair of electrons as a negatively charged "anion." To differentiate these, we will refer to the neutral ligands as "**L**" for short, and the negatively charged species as "**X**" (Figure 3.1b and c). We can envision the difference of these two ligand types using a molecular orbital diagram: in the case of an **L** ligand, both electrons are donated from the ligand orbital, while in the case of an **X** ligand, there is one electron contributed from the metal, and one from the ligand. It should be noted that this is purely a model: this kind of electron "accounting" makes it easier to differentiate whether bond cleavage results in ions (as is the case with **X**), or results in neutral species (as is the case with **L**).

This differentiation has an important implication for ALD: when an **L** ligand is lost, the resulting compound is neutral. Even if this bond breaking results in less thermodynamic stability for the metal-containing fragment, it can still exist in the gas phase. If an **X** ligand is lost, the resulting metal-containing fragment is a positively charged cation, and this is not stable in the gas phase (under reasonable conditions). So, **L** ligands can complicate the gas-phase chemistry of a precursor much more than an **X** ligand does.

One of the most common **L** ligands in organometallic chemistry is carbon monoxide, called a "carbonyl" when it is bonded to a metal center. This bonding is worth highlighting, since it involves a synergy of donating electrons from the ligand lone pair, while simultaneously accepting electrons from the metal in a pi bond (Figure 3.2).

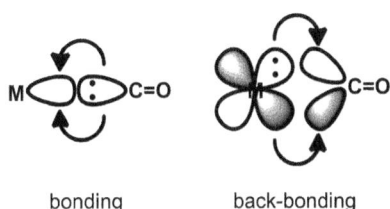

bonding   back-bonding

Figure 3.2: Back-bonding from a metal to a carbonyl ligand.

Back-bonding occurs due to a symmetric arrangement of empty orbitals on the carbonyl that matches full orbitals on the metal. This makes a "stronger-than-single" bond between the metal and the ligand. This comes at a cost to the carbonyl, since it is accepting electrons into an anti-bonding orbital, an orbital that destabilizes the structure of CO. This weakens the $C = O$ bond.

This type of bonding sets carbonyl ligands (and other ligands like them, e.g., isonitriles) apart since these are stronger than typical coordinative bonds, and the symmetry of the precursor can also influence how they react. The field of studying metal carbonyls is very well developed, but these organometallic species are relatively under-represented in ALD.

Another factor of ligand bonding that affects the stability of the resulting precursor is the "chelate effect." Chelation is a type of bonding where two or more bonds are made from the ligand to the central (metal) atom. In a molecule where the bond strength and electronegativities do not change, having a ligand bonding through more than one center offers stabilization by reducing the entropy of dissociation. Common ligands like the amidinates and beta-diketonates are examples of chelate ligands (Figure 3.3).

The stability of these ligands comes from the two bonds that they make to the metal center, but also because the structures have resonance: the bonding is symmetrical and so neither bond is fully **L** or fully **X**, but both are an average of these.

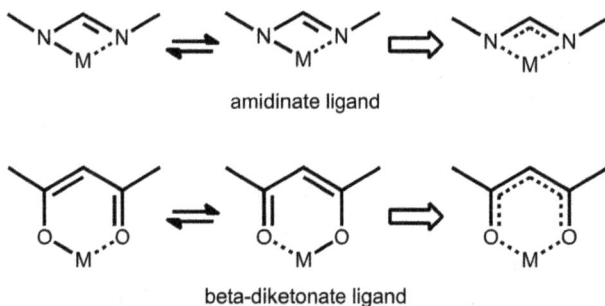

amidinate ligand

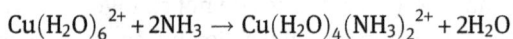

beta-diketonate ligand

Figure 3.3: Examples of chelate ligands used in ALD.

In resonance-stabilized chelates, the stabilization is greatly affected by the contribution of enthalpy: through the bond strength of the ligand to the metal center.

The chelate effect also provides additional stability by an entropic effect. When we consider loss of coordinative ligands in the gas phase, it is obvious that the loss of a chelate ligand (here, L_2) results in two particles in the gas phase, the original molecule, and the free ligand:

$$M(L_2)_n \rightleftharpoons ML_{n-2} + L_2$$

When there is a loss of the equivalent in "monodentate" ligands (i.e., a typical ligand L) results in three particles:

$$ML_{2n} \rightleftharpoons ML_{2n-2} + 2L$$

This example makes some assumptions, namely: that the M-L bond strength is the same in both examples, and that the particles do not interact in a meaningful way once the ligand has dissociated. Thus, entropy can be defined as the number of independent particles produced, and this highlights that a chelate L_2 ligand has less entropic drive to dissociate when compared to the similar loss of two L ligands in the gas phase. The chelate effect is best demonstrated by solution phase chemistry, where collecting the appropriate thermodynamic data is more straightforward.

If we consider a hexahydrated Cu^{2+} center, the addition of two coordinating nitrogen-containing ligand shows a difference in entropy. For the loss of two waters, and the addition of two ammonia ligands:

$$Cu(H_2O)_6^{2+} + 2NH_3 \rightarrow Cu(H_2O)_4(NH_3)_2^{2+} + 2H_2O$$

The enthalpy change (ΔH) from breaking and making the ligand bonds) is −46 kJ/mol, while the entropy change (ΔS) is −8.4 J•K/mol. The decrease in entropy here is due to the loss of available vibrational "modes" in free water compared to free ammonia. The free energy (ΔG) can be calculated from the standard definition for chemical free energy:

$$\Delta G = \Delta H - T \Delta S$$

For the above reaction, the free energy is −43 kJ/mol at 300 °C. This predicts that the reaction is thermodynamically spontaneous at this temperature.

When we look at the same reaction, but use ethylene diamine ("en," a linked, chelating ligand $H_2N\text{-}CH_2CH_2\text{-}NH_2$), we see a different contribution from entropy:

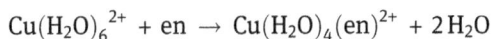

$$Cu(H_2O)_6{}^{2+} + en \rightarrow Cu(H_2O)_4(en)^{2+} + 2\,H_2O$$

The enthalpy change is relatively similar, −54 kJ/mol, but the enthalpy is quite different at + 23 J•K/mol. At 300 °C, the free energy is −61 kJ/mol. The chelate effect stabilizes the product $Cu(H_2O)_4(en)^{2+}$ better than the product $Cu(H_2O)_4(NH_3)_2{}^{2+}$.

Chelate ligands can also be used to provide steric protection (i.e., protection due to their size and the crowding they provide). One utility of these ligands is to enhance self-limiting saturation at the surface. Since these ligands are less likely to dissociate, they can be used at the surface to protect the payload atom from further reactivity. In this case, tuning of the metal-ligand interaction would be necessary to promote the retention of the ligand in the surface-bound moiety, but also to ensure that the ligand could be reacted away in subsequent pulses to provide the target film with minimal impurities. One significant cause of impurity inclusion in growing films is due to ligand retention, and so the atom that is directly bonded to the metal center is a very important factor to consider.

Metal-ligand bonds can easily be designed into a specific precursor, so ligands are often chosen for how easily their frameworks can be modified. A prime example of this is the family of ligands that include beta-diketonates, ketoiminates, and beta-diiminates (Figure 3.4a). Generally, the beta-diketonate can be made by enolization of the parent ketone with an acid chloride (Figure 3.4b). This can be modified by stepwise addition of a primary amine to replace the oxygen positions with nitrogen.

This is an interesting ligand since it is deprotonated at the carbon in the middle of the "bridge," and the resulting negative charge resonates and disperses between the two "E" atoms (where "E" is oxygen or nitrogen). This is a favorite ligand in vapor deposition (both CVD and ALD) due to its stability, as well as how easily modified it can be. Each "R" group (representing an alkyl moiety) depends on a different reactant, and so the ligand can have its symmetry and bulk easily modified.

The transformation of a beta-diketonate to include nitrogen chelating atoms does several things. First, the atom bonded to the metal center is different, and this will have a different bond strength. This can improve thermal stability and allow the ligand to remain with the metal center, in the case of stronger bonds, but can also make the ligand easier to remove in subsequent pulses, for weaker bonds. Since the family can be easily modified to allow all nitrogen bonds, all oxygen bonds, or an equal number of each, the bonding is highly tunable. This can also help control impurities left behind in a target film. For instance, a precursor being used to make a

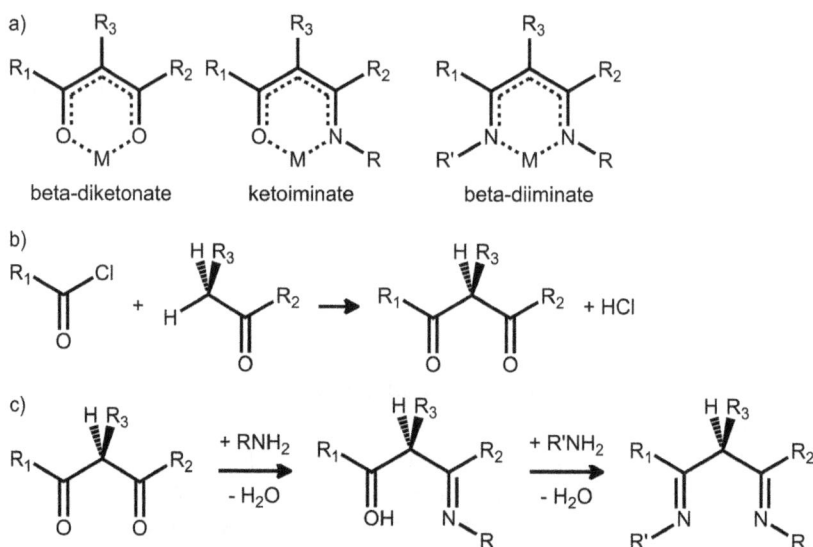

Figure 3.4: The family of ligands related to beta-diketonate, and their general syntheses.

nitride film should rather have metal-nitrogen bonds, since oxygen would be an impurity in these films. Finally, the addition of the nitrogen atoms brings steric bulk very close to the metal center, and the amount of steric influence that the R and R' groups exhibit can thus also influence bonding to the metal. If these groups are methyl (CH_3), this steric influence is small, and will allow the bond to be strong since the ligand can get closer to the metal center, increasing the overlap of the orbitals on the metal and the ligand that form the covalent bond. If these groups are tertiary butyl ($C(CH_3)_3$), the steric influence is large, and this will weaken the metal-ligand bond by forcing the ligand to be further from the metal center. Naturally, two chelating nitrogen atoms (in the case of the beta-diiminate) will weaken bonding more than one chelating nitrogen (for the ketoiminate) from doubling the steric hindrance.

The overall thermal stability will also be influenced by the choice of the atom bonded to the metal center. If we consider the first row of the d-block of the periodic table (the metals Sc, Ti, V, Cr, Mn, Fe, Co, Ni, Cu, and Zn), the electronegativities range from 1.36 to 1.91. Given the electronegativities of N (3.04) and O (3.44), these bonds can be quite polarized, with the electrons being withdrawn toward the ligand. If we apply Pauling's relationship between electronegativity and the ionicity of the bond, two things become apparent:

$$\% \text{ ionicity} = -4.94 \, (\Delta \chi)^3 + 23.0 \, (\Delta \chi)^2 + 5.75 \, (\Delta \chi) - 1.34$$

Here, $\Delta \chi$ represents the difference in electronegativities of the bonding partners. First, all M-O bonds have a higher ionic character than all M-N bonds due to the considerably higher electronegativity of oxygen. Among the metals being considered, scandium has the lowest electronegativity, which will lead to the worst mismatch in electronegativity with both nitrogen and oxygen. The mismatch with oxygen gives bond with 66% ionicity, showing a weakly covalent bond, while the similar Sc-N bond is 50% ionic. However, the Sc-O bond enthalpy is quite strong (682 kJ mol⁻) compared to the Sc-N bond (470 kJ mol⁻). This suggests that ligands making O-Sc bonds (like the well-studied tris (tetramethyl-3,5-heptanedionato)scandium(III), $Sc(thd)_3$) would retain its oxygen-containing ligands at intermediate reaction temperatures (due to the high bond enthalpy), but under chemical reaction, the ligand would more likely react through cleavage of the Sc-O bond (due to the higher ionic character of the bond). Contrarily, ligands making Sc-N bonds (like tris(diisopropylformamidinato)scandium(III), $Sc(^iPrAMD)_3$) would undergo thermolysis at lower temperatures and would have a higher likelihood of decomposing at a bond other than the Sc-N bond, and so could leave behind a nitrogen impurity. This analysis using bond enthalpies and electronegativities can be done on most metal-element combinations, and act as a first step in tuning the reactivity of the ligand in a potential ALD process. Of course, this is not infallible, but it can be a starting point for precursor design and ligand choice. In practice, process parameters and technical characteristics like the pulse length, second precursor, and carrier gas would also influence potential impurities. But during ligand design or selection, judicious choices can help alleviate issues (like impurity level) that appear in the final, deposited film.

A good, illustrative example of ligand influence is in the comparison of trimethylaluminum(III) ($AlMe_3$, TMA) and trichloroaluminum(III) ($AlCl_3$) in the deposition of aluminum oxide. Both precursors are of the type AlX_3, homoleptic (i.e., having one type of ligand) tricoordinate metal centers bearing a + 3 charge.

In general, halides are very volatile and thermally stable, and are often the first ligands considered in precursor development. However, chlorides (and halides in general) can have some significant disadvantages. Their by-products in a deposition process can be gas-phase Bronsted acids that is very corrosive to deposition equipment. Also, depending on the target film, halides can either passivate the surface by forming strong, non-reactive bonds to surface atoms or by etching the surface (forming a thermally stable and volatile halide species, as shown in Chapter 9). Chloride precursors are quite simple and inexpensive to produce.

Alkyls are more difficult to produce and purify, although $AlMe_3$ is an exception to this. Alkyls have a more varied volatility than halides but can also be very volatile at lower temperatures. Alkyls are also, in general, less thermally stable than halides, with weaker, but less ionic, bonds. This can lead to low-temperature decomposition that results in high carbon inclusion as an impurity of the deposited film. Methyl-containing precursors, specifically, can occupy a smaller volume than their chloride counterparts, which can lead to a higher GPC values by allowing a higher area density of surface-adsorbed metal-containing moieties.

Both $AlCl_3$ and $AlMe_3$ are dimers, existing under standard conditions as hexa-chloroaluminum(III) (Al_2Cl_6) and hexamethylaluminum(III) (Al_2Me_6) respectively (Figure 3.5).

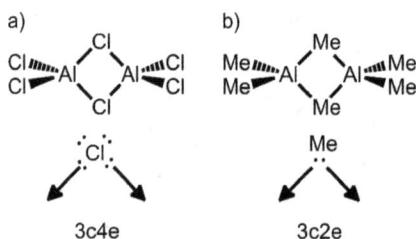

Figure 3.5: Bridging ligands in a) hexachloroaluminum(III) and b) hexamethylaluminum(III).

There is a significant difference in the way in which Al_2Cl_6 and Al_2Me_6 form dimers, and this is informative of bridging ligands in general. In the case of the chloride anion (Cl^-), each metal center is bonded through a pair of electrons, effectively making a 3-centered, 4-electron (3c4e) bond. In the case off the methyl anion (Me^-), the bridge is formed by a single electron pair, forming a 3-centered, 2-electron (3c2e) bond. Although this fundamental difference does not cause a difference in the reactivity of these two different precursors, a 3c2e bond is by its nature weaker than a 3c4e bond.

In the gas phase, both precursors are in equilibrium between their monomeric and dimeric states, and generally, the surface reaction of both occurs with the monomer. Generally, the reactivity of a precursor at a surface is considered to occur at a nucleation point, and so here a hydroxyl group is used to depict the surface defect on a growing aluminum oxide film's surface:

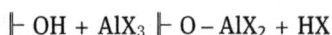

$$\vdash OH + AlX_3 \vdash O - AlX_2 + HX$$

The bond enthalpies and $\%_{ionicity}$ highlight that the alkyl would leave film impurities at lower temperatures. The bond enthalpy for Al-C is 268 kJ mol$^-$ and the $\%_{ionicity}$ is 20%. This is an intermediate bond with high covalency compared to Al-Cl, with a bond enthalpy is 502 kJ mol$^-$ and $\%_{ionicity}$ of 44%. Like the scandium example above, the weaker Al-C bond is less likely to ionize than the stronger Al-Cl bond, suggesting that $AlMe_3$ is more likely to leave a film impurity at low temperature than $AlCl_3$. In practice, alumina films from $AlCl_3$ are deposited using water or alcohols as the second precursor between 250 °C and 500 °C. The impurities, often including carbon when an alcohol is used, are low, with Cl lower than 2%, and often undetectable, and carbon lower than 2%. With alkyls, the thermal range of deposition can be from room temperature up to 300 °C (the thermal decomposition temperature of $AlMe_3$), also using water and alcohols. With this precursor, impurities of up to 30% carbon can be found at low deposition temperatures, with 3–5% more common near

250 °C. Although impurities can be limited by in-process tuning, as well as post-process treatment, the general trend highlights the importance of ligand selection: the necessary thermal range for deposition, the tolerance of the thin film's application to impurity, and the necessary robustness of the process need to be understood to choose an effective ligand for an ALD precursor.

Chapter 4
Precursors

Precursors for ALD can be divided into two general classes: precursors with metal atoms and those without. In most traditional processes, the metal containing precursor is referred to as the "precursor," and any non-metal precursor is commonly called the "reactant" or described by their role in the reaction. It is more rigorous to consider all process gases as precursors, and that language is used throughout this book. However, metal containing precursors share some common traits, and these will be highlighted below.

All gas phase reactants in an ALD process are precursors, and in this sense, all need to demonstrate self-limiting behavior. It has become more common recently for published ALD processes to show saturation curves for all gas phase reactants, and this greatly enriches the understanding of surface reactivity and mechanism. However, there are some key differences (some borne out of tradition) that need to be summarized. Commonly, precursors that bear metal atoms are considered the primary precursor, suggesting that there is a special importance to the precursor that carries what might be called the "payload" atom – the atom that defines the film, as aluminum does for aluminum oxide. It is also common to consider the non-metal reactant to define the chemistry of an ALD cycle: a metal containing precursor (like trimethylaluminum(III)) chemisorbs to a surface, and in an ideal approximation, does not undergo reaction until a second precursor is introduced. So the nature of the second precursor can often decide the target film, and so define the type of reaction.

For instance, ozone (O_3) is commonly called an "oxidant" because it provides oxygen to the film. However, it may or may not truly oxidize the metal center in the ALD process (i.e., removing valent electrons and increasing the metal center's positive charge). There are three different ways to look at ozone (as an example) as an oxidant precursor in ALD processes (Figure 4.1).

When ozone reacts with hexacarbonylmolybdenum(VI), the molybdenum starts with a 0 oxidation state, and ends with a + 6 oxidation state. Here, the metal center is clearly oxidized. The CO ligand could also be oxidized, depending on the conditions of the reaction. As shown in Figure 4.1, the CO, where carbon is in the + 2 oxidation state, is also oxidized to CO_2 with carbon in the + 4 oxidation state. This highlights the "ambiguity" of the word oxidant: although it formally means a species that, through its

Note: In this chapter, I would like to acknowledge my research group: what I know about ALD precursors comes in a large part from them. In particular, this chapter owes debts of gratitude to Dr. Jason Coyle and Dr. Matthew Griffiths. Through editing their PhD theses, my own understanding of precursors was challenged and expanded.

https://doi.org/10.1515/9783110712537-004

$$Mo(CO)_6 + 3\,O_3 \longrightarrow MoO_3 + 6\,CO_2$$

$$2\,AlCl_3 + 3\,O_3 \longrightarrow Al_2O_3 + 3\,Cl_2O_2$$

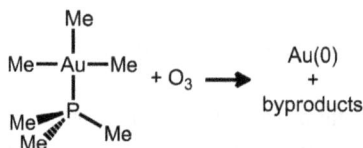

Figure 4.1: Ozone as an oxidant precursor in ALD.

own reduction will oxidize another center. However, this word sometimes gets used in the scientific literature to mean "adds oxygen to."

In the second reaction, where the reaction of trichloroaluminum(III) with ozone produces aluminum oxide, no formal oxidation occurs at the aluminum center: the aluminum is + 3 in the reactant as well as in the product. Although it is not common to report the side product of this reaction, the reasonable gas-phase by product is Cl_2O_2, a chlorinated peroxide that would be very reactive with oxygen and water (as well as pump oil, and other downstream chemicals it might encounter). In this structure, the chlorine atoms could be considered "oxidized," although formal charges are not a helpful or accurate way to think about this molecule.

In the third reaction, ozone is not considered to be reacting directly with the metal center, but rather is "combusting" the ligand system. The complicated reaction of combustion for this ligand system is not straightforward, the by products were not easy to characterize, and thus the formal reaction is difficult to determine accurately. However, the gold center starts in the + 3 oxidation state, and ends in the 0 oxidation state; this is a formal reduction. It is not at all accurate to say that the ozone reduced the gold center, but rather that gold, due to its strong noble chemical behavior, prefers a zero oxidation state, and so scavenged electrons during the complicated combustion step and reduced itself.

This ozone reactivity is meant to highlight the complicated nature of reactivity of precursors in ALD: these reactions tend, in the most part, to occur at a surface which participates in the reaction, and sometimes undergo a widely varied reactivity, with no specific reaction pathway being dominant. This requires precursor molecules to be relatively simple: since surface reactivity can run out of control along multiple reaction pathways, a smaller precursor with well-defined reactivity can allow for one reaction to emerge as the dominant pathway out of the general surface reactivity.

4.1 Atmospheric sensitivity

Metal containing precursors commonly react with atmosphere rapidly, making their handling and transfer difficult. This is on purpose, as the common reactivity of a precursor is to form a binary (or higher order) extended solid in an ALD process.

One of the most common reactions occurs between the metal containing precursor, with a ligand acting as a Bronsted acid, accepting a proton from either the surface, or a second precursor (which acts as a Bronsted base by losing a proton). Naturally then, these precursors commonly exhibit sensitivity to water. This characteristic is often necessary: a precursor that reacts strongly with water is very likely to react at a surface nucleation site with a dominant reaction:

$$MX_n + \vdash EH \rightarrow \vdash E - MX_{n-1} + HX$$

This very general surface reaction is meant to show that a volatile metal containing precursor (MX_n) with some number of anionic ligands will undergo a surface reaction at a nucleation site like an "$\vdash OH$" for an oxide surface or "$\vdash NH$" and "$\vdash NH_2$" for a nitride surface, for example. The driving force for this reaction is for the ligand to act a Brønsted base, abstracting a proton from the surface to make a conjugate (weak) acid. This is a fundamental surface reactivity in ALD and drives many processes forward. When designing a precursor for ALD, it is common to consider this reaction as the initial chemisorption step at the surface. This naturally leads to the predominance of halide-based and alkyl-based precursors. Other chemisorption steps will be discussed later in this chapter.

Less commonly, a precursor will react with atmospheric oxygen, or oxygen as a second precursor. In this case, there must be a reduction-oxidation (redox) reaction. A good example of this ids the deposition of ruthenium(IV) oxide from biscyclopentadienylruthenium(II):

$$Ru(CpEt)_2 + O_2 \rightarrow RuO_2 + side\,products$$

This general type of reaction can be thought of as a combustion reaction, and so the side products are difficult to predict. Depending on the extent of combustion, the ligand (which is comprised of only carbon and hydrogen, C_7H_9) might be oxidized all the way to their stable oxides: CO_2 in the case of carbon and H_2O in the case of water. Naturally, the combustion might not be complete, and produce CO, CH_xO_y, and a variety of other species. Generally, in ALD process development, the composition of the target film is paramount, and the characterization of the volatile side products is not common (and can be difficult). However, if we write the above equation as a compete combustion of the ruthenium(II) precursor to ruthenium (IV) oxide, it would be balanced like this:

$$2\,Ru(CpEt)_2 + 39\,O_2 \rightarrow 2\,RuO_2 + 28\,CO_2 + 18\,H_2O$$

This equation is balanced using the convention of not using a fraction in the stoichiometry and highlights how this combustion may not go to completion, considering 39 dioxygen atoms are required to fully combust one Ru^{2+} precursor. It also demonstrates that these types of redox reaction may produce a significant amount of water and promote hydroxyl formation at the film's surface, which may benefit the nucleation of further ALD cycles.

4.2 Halide precursors

The group 17 "halides" (i.e., F, Cl, Br, I, As) are very common **X** ligands in ALD precursors. Metal halide compounds are very thermally stable and simple to make. In general, they can be made from the metal or metal oxide and can be made pure in high abundance. They tend to be volatile, and so are often purified even in large scale by distillation or sublimation. The most common halide used is chloride (Cl⁻), and the primary example is trichloroaluminum(III) (AlCl₃, Figure 4.2).

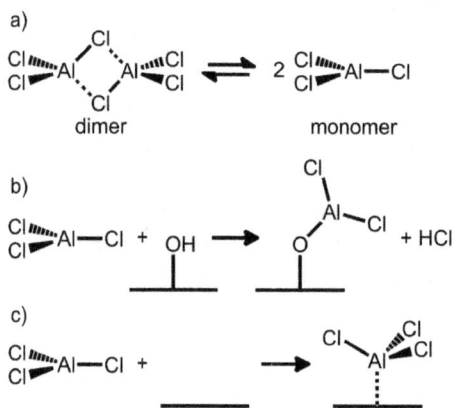

a)

dimer   monomer

b)

c)

Figure 4.2: The gas phase and potential surface reactivity of trichloroaluminum(III).

It is important to remember that ALD precursor can have gas phase reactivity, and in some cases, this reactivity is uncomplicated, but can still influence deposition. In the case of AlCl₃, this compound exists as a dimer, meaning that a chloride from each of two AlCl₃ molecules will bridge between the aluminum centers, making a coordination bond (shown dashed in Figure 4.2a). This can have a small effect on slowing down saturation, since saturation is more likely to be fast with the monomer, and slow with the dimer. In general, this type of equilibrium will shift to producing more monomer at higher temperatures due to an increase in entropy.

Two potential methods of surface adsorption highlight the reactivity of AlCl₃. The most common nucleation (as described above) is through acid/base reaction with a surface hydroxyl (Figure 4.2b). This is a very stable chemisorption, and loss of one **X** ligand means that the surface moiety cannot simply desorb; it would be a cation. This can be contrasted to physisorption of the AlCl₃ to the surface. As Figure 4.2a shows, the aluminum center wants to make four bonds. When it dissociates into a monomer, the aluminum center is a Lewis acid, requiring donation of electron density. If the surface is able donate some electron density to the metal, it could make a coordination bond (shown dashed in Figure 2c). Coordination bonds a generally weaker than covalent (molecular) bonds. The $AlCl_3$ physisorbed in this way can very easily revert into a neutral gas species and leave. This highlights the most significant difference between chemisorption (a more permanent, stronger surface interaction) and physisorption (a less permanent, weaker surface interaction).

4.3 Alkyl precursors

Alkyl-containing precursors cover a wide variety of precursors, and typically participate in Bronsted acid-base chemistry at surface nucleation sites:

$$MR_n + {\vdash}OH \rightarrow {\vdash}O - MR_{n-1} + RX$$

This fundamental reaction is formally a substitution reaction at the metal center, and for simplicity's sake can be thought to happen in one step at a surface. The thermodynamic driving force for this reaction is the formation of a very stable alkane (compared to the more reactive conjugate acid by-product produced by halides): although there is a risk of carbon contamination, this is less likely to occur from the alkyl ligand that is eliminated as an alkane and is more likely a problem caused by the alkyl groups left behind. Anecdotally, this reaction is the main reason that trimethylaluminum(III) is a successful precursor for TMA, and this process was deployed to a jewelry company in Helsinki in the early 2000s as the "nSilver" process by Beneq. The company wanted a solution to keep their silver jewelry from tarnishing on display, and so bought a simple ALD reactor to coat these silver pieces in alumina. To have the reactor accepted in shop, the company Beneq demonstrated to the local authorities that a typical process run produced less methane than a cow would in a 24-hour period. Now that is science.

The attractive nature of the stable, volatile product from this process is offset by two factors: alkyls are typically much more reactive to water than their halide counterparts, and alkyls typically show lower thermal stability. These two issues go together, in that processes employing alkyl precursors can generally be run at lower temperatures, but alkyls undergo complex and messy CVD-like deposition at higher temperatures, and so cannot be easily deployed for higher temperature processes.

4.4 General synthetic reactions

The synthesis of chemical compounds like ALD precursors is an extremely well-studied and active field, and there is no way to list all the potential synthetic reactions that one could use to make precursors. However, there are two fundamentally important synthetic reactions employed in the laboratory-scale synthesis of precursors: salt metathesis and alkane elimination.

Salt metathesis means "to exchange in the formation of a salt," and generally uses a metal halide as a starting material, and reacts it with a ligand base, resulting in the formation of a salt. Diethylzinc(II) is a common zinc-containing ALD precursor that can be made this way:

$$ZnCl_2 + 2\,LiEt \longrightarrow ZnEt_2 + 2\,LiCl$$

This fundamental reaction has an important thermodynamic driving force: the formation of lithium(I) chloride, the "salt" in the salt metathesis. When setting up such a reaction, the chemist will choose a solvent where the salt is sparingly soluble, or not soluble at all. This allows the salt to precipitate out of the solution (a thermodynamic system in this case), depleting one of the products of the reaction. Since the reaction is spontaneous, the system will undergo reaction to form more salt, and this ultimately ensures that the reaction is driven to completion. Naturally, this reaction can be performed with any ligand that can be made into a salt, and so is used in many precursor syntheses.

A second common reaction is the elimination of a volatile, stable gas, either from an alkyl precursor or a hydride precursor. By way of example, zinc(II) acetate is used as a precursor for the deposition of zinc(II) sulphide, and can be made from diethylzinc(II) and acetic acid (the acid that comprises vinegar, Figure 4.3).

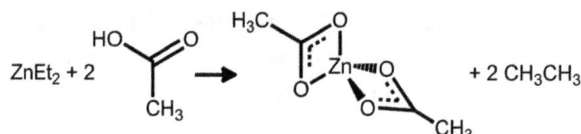

Figure 4.3: The alkane elimination of ethane gas from diethylzinc(II) to form diacetozinc(II).

Here, the driving force is like salt metathesis: the reaction results in a product that does not remain in the thermodynamic system (the solution) and so a spontaneous reaction will re-equilibrate the reaction continually, driving it to completion. Notably, the ligand that undergoes elimination here does not necessarily have to be an alkane. It can be any ligand that, when protonated, will for a volatile species that evaporates out of the solution.

In both cases, this is an example of Le Chatelier's Principle. A short definition of Le Chatelier's Principle is: a chemical reaction in equilibrium will respond to an outside condition that perturbs equilibrium be re-establishing equilibrium.

4.5 Designing precursors

Precursor design can have many requirements depending on application. Generally, easy to synthesize, scalable to large quantities, safe to handle, and inexpensive are the requirements for a precursor to be adopted in an industrial process. However, this does not get to the core of what characteristics are required of a precursor. For a precursor to undergo successful ALD deposition, it must be:

1. Volatile
2. Thermally stable
3. Reactive with the substrate and growing film surface
4. Self-limiting
5. Reactive to a second precursor

These characteristics can all be optimized individually, and so the art of precursor design is to creatively optimize each without sacrificing another. This design process usually involves designing, testing, and redesigning precursors until they are acceptable for a given application. Other concerns can include tuning the melting point, controlling impurities in the resulting film, and others. Naturally, the list can be extensive, and differs from process to process.

4.6 Volatility

Volatility has two characteristics that are important for precursor development: the precursor needs to be volatile at low temperature, and the precursor should volatilize at a high rate to minimize pulse length.

One prevalent idea in volatility is that the molecular weight has a role in volatility. This concept comes from Knudsen's derivation of effusion of a volatile species through a hole of a known diameter:

$$p_v = \frac{\Delta m}{A_o \Delta t} \left(\frac{2\pi RT}{M}\right)^{\frac{1}{2}}$$

This formulation of the Knudsen method is derived from the Kinetic Theory of Gases and shows that vapor pressure (p_v) is inversely proportional to the square root of the molecular weight (M). This relationship itself comes from a derivation of the speed of a particle in the gas phase, and has, at its heart, an assumption that the particles in the gas phase behave in an ideal fashion and follow the Ideal Gas Law ($pV = nRT$). Thus, Knudsen effusion assumes that there are no interactions between molecules.

This is more reasonable for the gas phase than it might seem on first consideration: when a system has a very low pressure, gas particles are, on average, far enough away from each other that interaction does not affect their behavior. The rare close interaction of particles does not affect the system enough to be noticed. However, this assumption is less valid in the condensed phase (either as a liquid or solid precursor in a bubbler). Here, interparticle interactions can be strong and plentiful, and cannot be ignored. Indeed, there can be a tendency for intermolecular interaction to lessen as molecular weight lessens, simply because there are fewer ways for particles to interact with each other.

Generally, the thermodynamics of evaporation (or sublimation, in the case of solids) control volatility, making this process always endothermic, i.e., heat is always required by the molecule. From a thermodynamic standpoint, this means that the enthalpy (ΔH) will be negative. Thus, for volatility to be spontaneous, the entropy (ΔS) of volatilization must positive (i.e., the system needs to become more disordered). This is obvious when considering the common definition of Gibb's free energy (ΔG), the energetic predictor of thermodynamic spontaneity. A thermodynamically spontaneous reaction has a negative ΔG:

$$\Delta G = \Delta H - T\Delta S$$

In most thermodynamic conditions, the entropy is about three orders of magnitude less influential than the enthalpy. That is to say, the energy associated with entropy is 1/1000 the enthalpy, in a general sense. It becomes obvious from the simple definition of Gibb's free energy that volatility would occur when the temperature is high enough to equilibrate enthalpy and entropy. At its core, the optimization of volatility requires the lowest possible enthalpy of volatilization and the highest possible entropy of volatilization.

The enthalpy of volatilization can be controlled by limiting the interactions between individual molecules. The common interparticle interactions are electrostatic, hydrogen bonding, dipole-dipole, ion-dipole, and van der Waals interactions. These can be controlled by the shape and bonding in a precursor. For instance, a polar bond, where the difference in electronegativities for the atoms is large, will allow greater interaction between individual molecules and suppress volatilization. Aluminum(III) chloride is less volatile than trimethylaluminum(III) for this reason. The electronegativity of aluminum is 1.61, and so the difference is large with chloride (EN = 3.16, Δ_{EN} = 1.55) than it is with carbon (EN = 2.55, Δ_{EN} = 0.94). Although there are other factors that affect the interaction of molecules, electronegativity is a good starting point for the minimization of dipole interactions. Ligands that have hydrocarbon groups tend to help volatility since they are non-polar, and so can protect a precursor molecule from interactions that would prevent it from volatilizing. Another very effective atom for increasing volatility is fluorine: it is not easily polarized, and it keeps its electrons tightly held so there is minimal interactions with other groups. A good example of this is the series of betadiketonate dimethylgold(III) compounds, which show increasing volatility with increasing inclusion of fluorine (Figure 4.4).

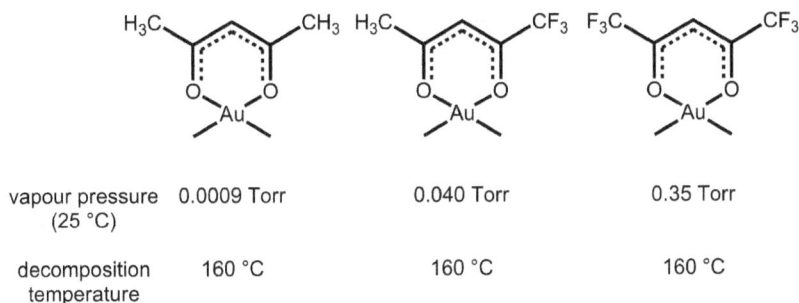

vapour pressure (25 °C)	0.0009 Torr	0.040 Torr	0.35 Torr
decomposition temperature	160 °C	160 °C	160 °C

Figure 4.4: Increasing volatility of related gold precursors with increasing CF_3 inclusion.

These gold(III) precursors are all based on the same ligand framework, and very likely share a decomposition mechanism, considering that they all decompose at the same temperature. However, the volatility of this compound can be increased by the inclusion of fluorine groups (in this case, replacing methyl, CH_3, with trifluoromethyl, CF_3). The less polarizable nature of fluorine compared to hydrogen limits the intermolecular interactions enough that there is a measurable change in volatility at 25 °C. One downside to this design concept is that the opportunity to have fluorine impurities in a film increases with increased fluorine content. If a target film's application is intolerant of fluorine inclusion, this concept might not then be used.

4.7 Thermal stability

Thermal stability is a difficult design parameter to tune for precursors: it relies on strong, stable chemical bonds between the ligands and the payload atom in the precursor. The specific chemistry of the payload atom and its bonding partner dictates this stability. For instance, electropositive metal centers (e.g., Al, Zn, Ti) make strong and stable bonds to oxygen, and so ligands using this heteroatom can be very stable. This naturally makes oxygen a common impurity in films that employ electropositive elements and oxygen-containing ligand systems, as well as making it very difficult to deposit metallic films of these elements. However, electronegative elements (e.g., Au, Pt, Pd) do not make strong bonds to oxygen, and oxygen-containing ligand systems tend to make precursors from these metals thermally unstable. Likewise, it is almost impossible to make oxide films using these metals, and even using water, oxygen, or ozone as second precursors in processes using these atoms will result in metallic films.

Design of thermal stability in precursors tends to follow two paths: to add "chelate ligands" when possible, and to design out low temperature decomposition pathways. This often requires iteration of design and testing, and a precise mechanistic understanding of how a precursor decomposes thermally. Below are a few common precursors and their thermal decomposition pitfalls, with a discussion of how to alleviate them.

4.7.1 The chelate effect

In a high-valency precursor, it is common to use a homoleptic approach: to employ the same ligand, bonded through the same element to fill the valence of the central atom. Titanium(IV) is a common metal center for the deposition of TiO_2. When tetraisopropxotinamium(IV) (**TTIP**, Figure 4.5) is used as a precursor, the shift between ALD growth and CVD growth – demonstrating surface decomposition of the precursor – occurs at 240 °C. Using a similar precursor, where all of the bonds to the Ti(IV) center are still oxygen, but two of the ligands form a ring to the metal center (bistetramethylheptanedionatodiisopropoxotitanium(IV), **chelate**, Figure 4.5), the surface decomposition starts 120 °C higher in temperature, at 380 °C. This is an example of the "chelate effect," where ligands that bond to the metal center through more than one group show enhanced stability when compared to ligands that do not.

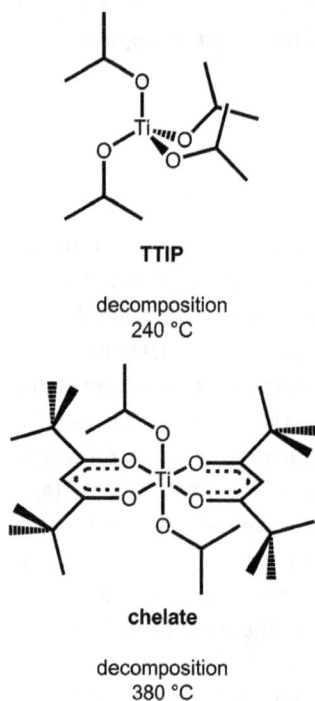

TTIP

decomposition
240 °C

chelate

decomposition
380 °C

Figure 4.5: The structures of two Ti(IV) oxygen-containing precursors showing different surface decomposition temperatures due in part to the chelate effect.

The chelate effect is an entropic effect: the decomposition to lose several ligands (thus increasing disorder) requires less energy overall to occur, and so can occur at lower temperatures. When a chelate ligand is lost, the system is less disordered, and this generally results in a higher temperature to become stable. Thus, precursors that suffer from low thermal decomposition temperatures can be improved by replacing some of the monodentate ligands with multidentate ligands. The thermodynamics of chelation are discussed in further depth in the previous chapter.

4.7.2 Low temperature reactivity

One aspect of precursor design that can drastically improve precursor performance is the identification of low-temperature decomposition pathways. This requires an in-depth knowledge of the decomposition mechanism of a precursor, and so requires a significant amount of laboratory work. However, thermal stability can be drastically improved.

A good example of this is the improved behavior of gold(I) nitrogen-based chelate ligands. Early attempts to deposit gold metal by ALD used N,N-dimethyl-N'-N'-diisopropylguanidinatogold(I) (Figure 4.6). This compound underwent CVD growth at 150 °C and had a tremendously high growth rate (39 nm/min), making it useless for gold metal ALD. The thermal decomposition followed two main pathways: the ligand would eliminate carbodiimide to produce a thermally unstable amide, as well as undergoing a beta-hydrogen elimination to produce a thermally unstable hydride. Beta-hydrogen elimination is a reaction where a hydrogen bonded to a center that is two bonds away from the metal center is abstracted by the metal center, and it is a common inorganic reaction pathway.

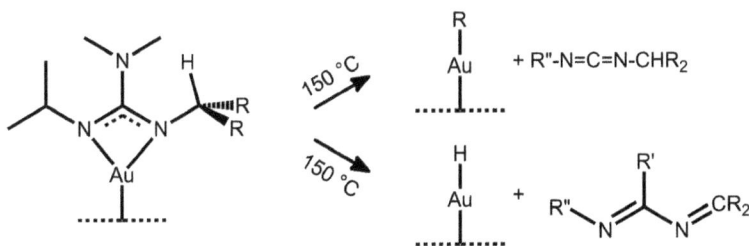

Figure 4.6: Surface decomposition pathways for a gold(I) guanidinate precursor.

A redesign of this precursor needs to solve both problems: this redesign resulted in a pyrrole-based ligand (Figure 4.7). The pyrrole ring made it much more energetically costly for the ligand to undergo the elimination of the carbodiimide, and all beta positions were occupied by methyl groups, alleviating the low-temperature beta-hydrogen elimination. The subsequent thermal decomposition did not start until 275 °C, showing an improvement of 125 °C. Indeed, this precursor did not undergo CVD deposition of gold metal until 350 °C, and then the growth rate was only 1 nm/min.

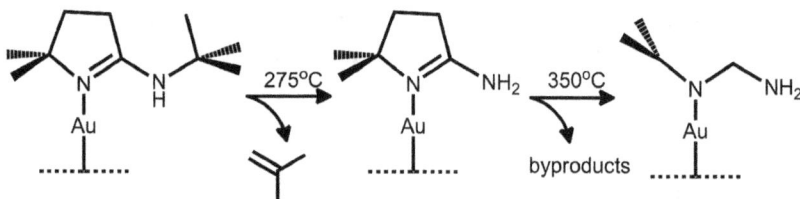

Figure 4.7: Improved thermal stability was found with a gold(I) precursor using a pyrrole-based ligand.

It should be noted that this specific precursor redesign, although successful, required the synthesis of a ligand that would be too costly to make on an industrial scale. This highlights another design rule: keep it simple.

4.7.3 Melting point

Although making a precursor with a low melting point is not a necessary design element, a low melting point can be useful for process reproducibility. When volatilizing a precursor, it is better that the precursor be a liquid: the surface area is more uniform, and if impurities exist at the surface, they are mobile, and have little influence. Both aspects improve the kinetics of volatilization and help keep a precursor delivering a constant vapor pressure of precursor. In a solid, the surface area that volatilization can occur from is defined by the surface area of the particles. As the process continues, the particles get smaller, and so the surface area decreases. If the volatilization is kinetically slow (i.e., similar to the length of the pulse in the ALD cycle), this could result in a lower pressure of precursor delivered later in the process, compared to early in the process. Likewise, if impurities accumulate on a solid precursor surface, they are les mobile than they would be on a liquid, and thus decrease the surface area further.

However, a solid material is more highly desired for shipping and handling. The same surface area argument that helps volatilization will make the precursor kinetically more reactive, and so a spilled liquid can be more dangerous than a spilled solid. As well, a sold powder is easier to handle (in general) when filling a bubbler. An ideal melting point for a precursor would be above the temperature of shipping and handling, but below the process temperature.

Selecting a melting point for a precursor is not possible but tuning the existing melting point for a precursor can be surprisingly effective. Generally, the melting point can be lowered by the inclusion of long, unbranched alkyl chains extending from the ligands. An excellent example of this is the modification of hexacarbonyl-tungsten(0) ($W(CO)_6$) where one carbonyl is replaced by an isonitrile with a variable chain length (Figure 4.8). The symmetric distribution of electron density in $W(CO)_6$ means that intermolecular interactions that keep it in the condensed phase are relatively uniform and distributed over its highly symmetric octahedral coordination geometry. By disrupting this symmetry on one vertex and adding an alkyl chain in this position, the melting point could be easily lowered.

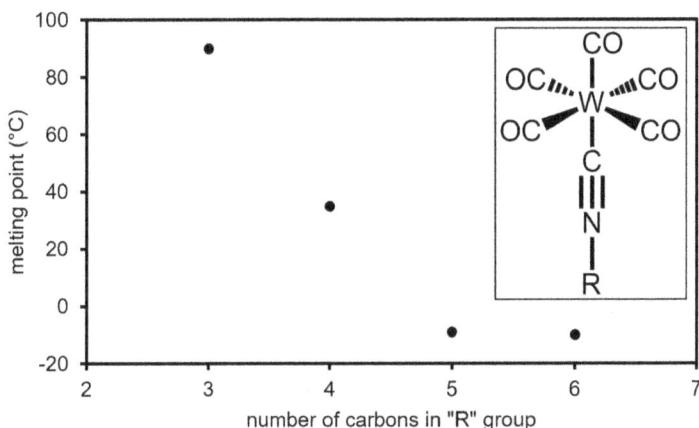

Figure 4.8: Control of melting point by precursor design.

In this case, the melting point can be drastically altered by precursor design. $W(CO)_6$ has a melting point of 170 °C, and the replacement of on of the "CO" groups with an isonitrile (CNR) can drop this to below 100 °C:

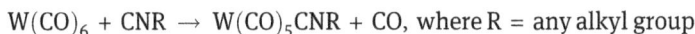

$$W(CO)_6 + CNR \rightarrow W(CO)_5CNR + CO, \text{ where R} = \text{any alkyl group}$$

As the alkyl chain gets longer ($C_3 - C_5$), the melting point drops linearly to below 0 °C, suggesting that a butyl (i.e., 4-carbon) alkyl chain would give useful melting point: above room temperature so that it could be handled and shipped as a solid, but low enough to melt to a liquid for the deposition process. This phenomenon is attributed to the "frustration of crystallization," where the alkyl chains can adopt a variety of different conformations, making it difficult to have them adopt the same orientation as lattice points in a crystal. This allows greater mobility at lower temperatures and causes a lowered melting point. Interestingly, there is no gain after C_5, and a hexyl (6-carbon) chain has a similar melting point to the pentyl (5-carbon) chain. Indeed, with an 8-carbon chain, the melting point goes back up to ~ 20 °C, suggesting a second phenomenon comes into play. This design ploy has been used in a variety of precursors to tune melting point.

4.8 Reactivity

Building reactivity into a precursor is a balancing act and requires creativity: the precursor must be reactive with a surface, and with other chemical precursors without being thermally reactive. To put this another way, the precursor must exhibit *intermolecular reactivity* without exhibiting *intramolecular reactivity*. Although reactivity (particularly with a second precursor) can be specific to a precursor and a process, many precursors can be tuned in the same way to react with the surface or growing film. By including a specific ligand in the precursor design that will react at a surface nucleation site, saturation can be tuned. Generally, not all ligands in the precursor should be tuned to react at a surface, and so a precursor with a variety of ligands can be advantageous to control ALD. Such chemical species, with different types of ligands coordinated to the same metal centers, are called heteroleptic species. In general, this type of precursor will be designed with a ligand that will react to chemisorb the precursor to the growth surface, as well as a ligand that will remain bonded to the metal and protect it from reacting with other precursors in the gas phase.

To consider what would make a good anchor ligand, the pK_a of the ligand species should be considered. The K_a of an acid compound is the dissociation equilibrium constant of that acid, according to the following chemical reaction:

$$RH \rightarrow R^- + H^+$$

$$K_a = \frac{[H^+][R^-]}{[RH]}$$

The pK_a of a chemical species is a definition of how easily it reacts to give up a proton and is based on the well-known definition of pH, but rather than just being concerned with water (like pH is), pK_a is concerned generally with all acids:

$$pH = -\log[H^+]$$

$$pK_a = -\log(K_a)$$

This definition is specific so that it can be easily used with pH to predict the behavior of an acid in water, but we can adapt these definitions for ligand reactivity. Here, we can consider the ligand to be the species R^- (i.e., the base) and RH to be the ligand reacted with a surface proton (i.e., the conjugate acid):

$$MR_n + {\vdash}OH \rightarrow {\vdash}O - MR_{n-1} + RH$$

Due to the negative sign and the logarithmic nature of the definition, the value of pK_a will be an increasingly negative number as R^- is able to release H^+. To put it another way: the weaker R is as a base, the smaller the value of pK_a. However, ligands need to bond strongly to a metal center, if our surface reaction produces RH, we want R^- to also bond to H^+ strongly. This means that R^- needs to be a strong base, and stronger bases have larger pK_a values. The table below shows some typical ligands and their pK_a values.

Table 4.1: Typical ligands in ALD, and their pK_a values, and their type of bonding. Note that ligands with a variable "R" group will have a variety of pK_a values depending on the definition or R.

Ligand	pK_a	Ligand	pK_a	Ligand	pK_a
Chloride, Cl^-	−8	Betadiketonate	~9	Amide, R_2N^-	~11
Amidinate	~12	Thiol, RS^-	~13	Cyclopentadienyl, Cp^-	16
Alkoxide, RO^-	~17	Hydride, H^-	~35	Alkyl, R^-	~50

Although it is impossible to list every ligand used in ALD precursors, Table 4.1 contains the pK_a values for a variety of different common ligands. A higher pK_a means that it is more difficult to form the ligand from its protonated form.

By pK_a analysis, it is easy to see that a precursor that has both chloride and alkyl ligands – for example dichloromethylaluminum(III)) – would react with a surface hydroxyl group through its methyl group, leaving the chloride groups in the chemisorbed precursor (Figure 4.9).

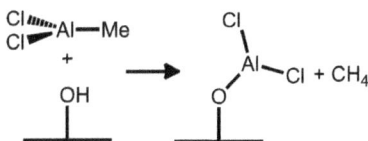

Figure 4.9: The preferential reaction of the alkyl group Me^- over the chloride groups Cl^- at a surface hydroxyl.

This simple example demonstrates an important design tool for precursors: by carefully considering what type of ligand will preferentially react at a surface, the chemisorbed surface moiety can be selected. This further enhances reactivity control by allowing specific ligands to be left over to participate in the further deposition mechanism of the target film by ALD.

Chapter 5
Thermolysis

An ALD precursor can be expected to undergo different chemical transformations at a surface depending on the target film and, therefore the choice of other precursors in the process. However, a successful ALD precursor needs to be volatile at a useful temperature, and non-reactive to intramolecular reactions in this range, and this is independent of the intended ALD process or target film. That is to say: there can be no ALD process if a precursor is not volatile nor thermally stable. This makes these two characteristics of a precursor of paramount importance.

It is an interesting exercise to measure both volatility and thermal reactivity, and attribute these to a temperature: both volatilization and thermal decomposition are chemical reactions with specific kinetic mechanisms, and there is no "on-off" temperature for them. They may become thermodynamically spontaneous at a given temperature, but as temperature increases, each can become faster, and therefore more impactful over the timeframe of an ALD cycle. This begs the question: what are acceptable definitions for the "onset" of volatility and thermal decomposition? When are these strong enough effects to affect a process? Naturally, the definition of each can be entirely process dependent.

5.1 Definitions of thermal onset temperatures

Two useful definitions for the thermal characteristics of a precursor are T_V (the volatilization temperature), the lowest temperature where the precursor produces enough vapor pressure to enable a process, and T_D (the decomposition temperature), the highest temperature where thermal decomposition can be ignored. A common definition for T_V is the temperature where a precursor produces one Torr (which is 0.00131579 atm, or 133 Pa) of vapor pressure. This is useful in flow reactors, since carrier gases often have pressures in the 1–5 Torr range, and so the precursor is not overly diluted by the carrier gas. For T_D, there is much less of a straight-forward definition, but it should be highlighted here that the decomposition in question is the decomposition of the precursor before, or as, it volatilizes, and so this temperature can be thought of as the temperature at which the precursor is kept in the tool; the "bubbler temperature." This chapter will use T_V and T_D to denote these thermal characteristics without further worrying about their specific definitions.

Note: In this chapter, I would like to acknowledge Prof. Arun M. Umarji, whom I have never met. Prof. Umarji's group published a paper in 2008 about thermally evaluating precursors for CVD, and I think about this paper all the time. I will buy him a beer if I ever meet him.

https://doi.org/10.1515/9783110712537-005

5.2 Thermogravimetry

The measurement of thermal characteristics of many chemical compounds can be successfully carried out by thermogravimetry: a characterization technique that monitors the mass of a sample during a programmed temperature treatment. Generally, this is performed on a specialized instrument equipped with a microbalance for mass measurement, as well as a well-controlled furnace for programmed temperature profiles (Figure 5.1).

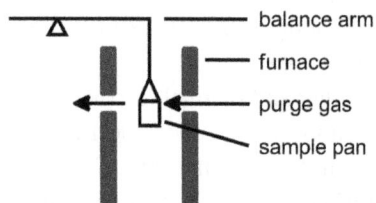

balance arm

furnace

purge gas

sample pan

Figure 5.1: A typical thermogravimetric analysis setup.

The typical thermogravimetric experiment that is run to test a potential precursor for ALD is a ramped temperature experiment: the pan is loaded with a sample (typically between 10 and 50 mg) and the temperature is ramped at a constant rate (typically 10 °C/min). This serves to easily show whether a sample will volatilize, or whether it has low temperature thermolysis pathways that prevent it from going into the gas phase intact. Volatility is characterized by three main characteristics: there is one thermal event denoted by the lack of inflections in the mass loss, the derivative of the mass loss curve shows an exponential onset, and there is zero residual mass (Figure 5.2).

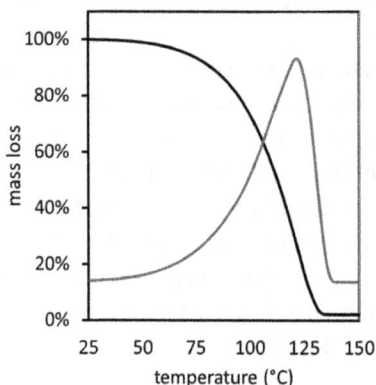

Figure 5.2: A typical ramped thermal experiment showing volatility of the sample. The black line is the mass loss with respect to temperature, and the grey line is the derivative of the black line.

Here, the black line shows the mass loss of the sample as a percentage as the temperature is ramped. This line shows a smooth and increasing mass loss with respect to temperature, uninterrupted by any other thermal feature. This is highlighted in the derivative curve (grey line) where there is an exponential onset until the mass is entirely lost. At this point, the derivative very quickly drops to zero. Finally, the mass loss curve plateaus very close to zero, demonstrating that all the mass is lost in this single thermal event. Although this is not conclusive evidence, a thermogravimetric trace with this form is very encouraging in precursor development. Kinetics can be probed in a ramp experiment by varying the thermal ramp rate. In literature, the most common ramp rate is 10 °C/min, which is a common factory setting in thermogravimetry instruments. By setting the ramp rate lower, the sample mass will reside longer at each temperature and so any thermal degradation that occurs over the length of the TGA experiment will be highlighted, and if there is a slow decomposition, this will result in a higher residual mass. Naturally, if there is no decomposition event, the TGA trace will again resemble Figure 5.2. The difference will be the temperature at which mass loss is first noticed. Since the ramp rate is slower, more mass will be lost at each temperature, and so the thermal "onset" temperature will shift to slightly lower temperatures for lower mass loadings, and the drop in mass will be much steeper (Figure 5.3). This is a good reason to not wholly rely on the noticeable thermal onset by TGA: it will be varied by experimental conditions, including ramp rate and sample mass.

The variability in "thermal onset" by sample mass can be exploited to further interrogate thermal stability. This further set of experiments can be called the "thermal stress test." This test involves several different traces using the same ramp rate, but altering the sample mass: in this way, higher sample masses will have more sample in the pan at higher temperatures, since the rate of volatilization is constant. This can stress a larger mass of sample at higher temperatures, and a small change in residual mass between experiments demonstrates a slow thermal decomposition (Figure 5.3).

This simple experiment shows that the precursor in question has a slow thermal decomposition. Under lower mass loading, it is hardly noticeable, but as a higher absolute value of mass is subjected to higher temperatures, the residual mass increases from 2% at ~ 10 mg to 14% at ~ 60 g. This type of thermal degradation would be problematic if the precursor were kept at a moderate temperature of extended periods of time: perhaps days or weeks. Thus, a thermal stress test can highlight precursors with a significant risk of decomposition in the bubbler of an ALD tool.

Naturally, TGA can be run at a constant temperature, and produce an isothermal mass loss. This experiment can be used to test a precursor's rate of delivery at a given temperature. It should be noted, however, that a TGA with a purge gas might give a higher reading than a system where the vapor develops in a constant headspace. With a purge gas, there is no vapor pressure over the volatilizing precursor:

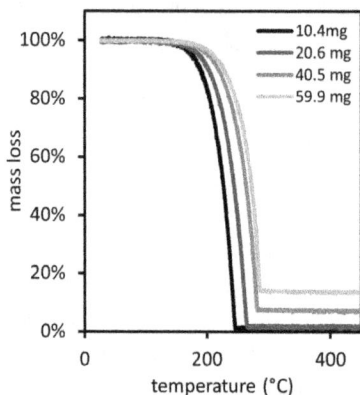

Figure 5.3: 10 °C/min TGA ramp experiments with variable mass loadings.

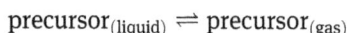

$$precursor_{(liquid)} \rightleftharpoons precursor_{(gas)}$$

When we consider evaporation as a chemical equilibrium, having the purge gas remove vaporous precursor will promote quicker evaporation by Le Chatelier's principle. In a "static" system, the evaporation rate would slow down as the vapor pressure built up, and this system would reach equilibrium. This is the main issue with measuring volatilization and mass transport by TGA: the dynamic nature of the purge system favors quicker volatilization and so may overestimate thermodynamic vapor pressure. However, the purge used in TGA conveniently emulates the removal of the vapor in an ALD process during the pulse, and in practice, the estimation of vapor pressure from TGA data is useful and insightful, particularly to compare the thermal characteristics of a set of similar precursors.

A careful measurement of vapor pressure using TGA requires an experiment called a "stepped isotherm," where the temperature ramp program increases by a standard increment, and then rests at that temperature long enough for the precursor to stabilize at an isothermal mass loss (Figure 5.4).

Here, 5 °C increments (grey trace) are each stabilized for 7 min, and a mass loss is recorded. Each step is a 7-minute isotherm, and the mass loss is constant over this period (although these linear mass losses are difficult to see in Figure 5.4). Thus, each step produces a value of $\Delta m/\Delta t$, which can be used to calculate the vapor pressure.

Vapor pressure can be calculated from the Langmuir vapor pressure equation:

$$p_{vap} = \frac{\partial m}{\partial t}\sqrt{\frac{2\pi RT}{\alpha_1 M}} = \frac{\partial m}{\partial t}\sqrt{\frac{2\pi R}{\alpha_1}}\sqrt{\frac{T}{M}} \approx k\frac{\Delta m}{\Delta t}\sqrt{\frac{T}{M}}$$

This equation is derived from the Kinetic Theory of Gases, where M is the molecular weight of the precursor, T is the temperature of the isotherm, and R is the ideal gas constant.

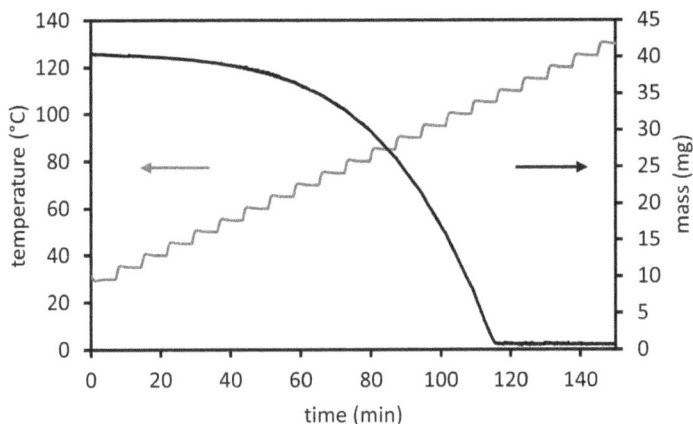

Figure 5.4: A stepped isotherm curve used for the determination of vapor pressure by TGA. Here, the grey trace represents the temperature, and the black trace represents mass loss.

As with previous examples where derivations come from the kinetic theory of gases, this definition assumes that the gases act ideally, and do not have any intermolecular interactions. The variable α_1 is used to correct for these ideal assumptions, and in vacuum is set to 1. Although this is generally valid, in some cases this could be a source of error in vapor phase estimation. Note that the increment in the above equation is $\partial m / \partial t$, and so a further assumption that we will make is that his term can be used macroscopically as the experimentally determined $\Delta m / \Delta t$.

When vapor pressure is determined from TGA, the value α_1 can be experimentally determined, and controls for all experimental, systematic errors, as well as for the change in thermodynamics from a vacuum system. Generally, a constant of k is determined for all the non-experimental variables, using a standard where the vapor pressure is known for a wide temperature range using a second measurement technique.

Having used the Langmuir vapor pressure equation to estimate vapor pressure at each temperature, a linear plot of pressure and temperature can be generated. This requires the use of the Clausius-Calpeyron equation:

$$\ln p_{vap} = A - \frac{\Delta H}{RT}$$

A plot of $\ln(p_{vap})$ vs. $1/T$ can reveal two beneficial thermal parameters (Figure 5.5). It should be noted that temperature needs to be converted to Kelvin, since the "zero point" in Celsius will cause problems in this graph.

This curve can be used to determine T_V, which was defined earlier to be the temperature at which one Torr of vapor pressure is produced. From the Clausius-Clapeyron equation, it should also be obvious that the slope of this line is ΔH_{vap}. Although the number of assumptions that have been made to generate these data might make this value inaccurate, it is very helpful when comparing two or more

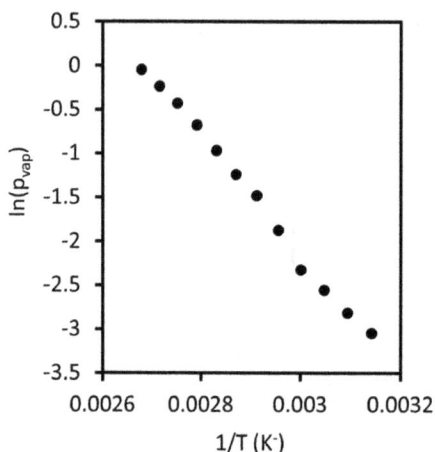

Figure 5.5: The Clausius-Clapeyron vapor pressure curve generated from the data shown in Figure 5.4.

precursors. If the vapor pressure lines show the same slope, in general these precursors share the same volatilization thermodynamics, and so have similar intermolecular forces as well as having similar kinetics of volatility. If a family of precursors are compared, this is a good method to identify any precursors that exhibit unusual thermal behavior.

If the kinetics of volatilization is reasonably fast, then a stepped isotherm is not necessary to generate the above Clausius-Clapeyron curve: each data point in a standard ramp experiment can be used with the previous point to determine an instantaneous $\Delta m/\Delta t$ term: the tangent to the slope of the volatilization curve at that point. This can provide a quick way to determine an estimate of T_V (albeit with increases error), and a stepped isotherm experiment can be subsequently run to corroborate these data when necessary.

5.3 Differential scanning calorimetry

As mentioned previously, DSC can be used to interrogate the thermal stability of a precursor. Like using TGA, this will not give an exact value, and even the definition of T_D can be debated as to what it should mean. However, TGA only gives decomposition information if decomposition occurs before or during the volatilization of a precursor. For volatile precursors that show no residual mass nor any slow decomposition by a thermal stress test, DSC is a simple method to understand at what temperature thermal decomposition pathways could interfere with a precursor's ability to form self-limiting monolayers at a surface. In this experiment, the DSC sample pan must be sealed, without a pinhole to alleviate pressure. If the sample mass can be lost in this manner, the DSC will only give information comparable to the TGA.

A DSC instrument is an elegant method of measuring thermodynamic events over a range of temperatures (Figure 5.6).

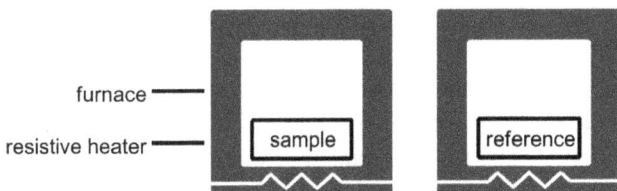

Figure 5.6: The principle of differential scanning calorimetry.

In DSC, a sample and a reference are resistively heated over a temperature range. The temperature is monitored and kept constant between the sample and the reference. The reference is chosen so that it does not undergo any thermodynamic event (melting, boiling, decomposition) over the available temperature range. If the sample does, power is either increased or decreased in the resistive heater to ensure that the temperature between the two furnaces stays the same. In this way, if the sample undergoes an endothermal event (i.e., that requires energy, like melting), the power will be increased to the sample heater. If the compound undergoes an exothermal event (that releases energy, like thermal decomposition), the power will be decreased to the reference heater. These changes in power can be plotted against temperature to highlight the thermodynamic events experienced by a sample during a given temperature change (Figure 5.7).

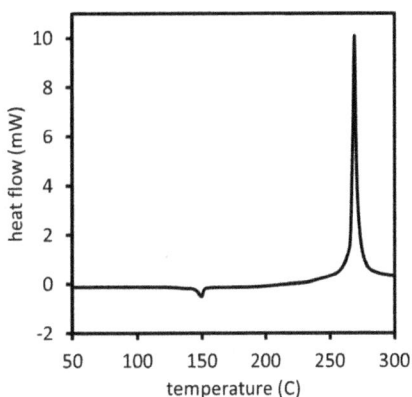

Figure 5.7: A differential scanning calorimetry trace showing an endotherm at 150 °C and an exotherm at 270 °C.

The two types of thermodynamic events are opposite with respect to energy flux and are typically graphed together with endothermal events oriented down and exothermal events oriented up. In the example in Figure 5.7, the very strong exotherm at 270 °C is likely a thermal decomposition, while the endotherm at 150 °C is likely due to melting. The precursor may need to be independently tested at the higher temperature to verify this decomposition and indeed, such verification can help determine the mechanism of decomposition. Like TGA, DSC can give a rapid assessment of a precursor's thermal characteristics to help determine whether further investigation is worthwhile.

Another technique that is commonly employed in DSC is to cycle through the temperature range two or more times to test the repeatability of the thermodynamic features. This has the benefit of differentiating events that will reoccur (like a melting endotherm) from those that will not (like a decomposition exotherm). Naturally, these cycling experiments must be planned judiciously to highlight which peaks are reversible, but this can aid in defining T_D.

5.4 Thermal range and figure of merit

To select a precursor for a given target film or ALD process, two of the most important characteristics to consider are the thermal onset of volatility T_V and the thermal onset of decomposition T_D. These give a thermal range in which an ALD process can be sought: when the precursor has a high enough vapor pressure to be used in a process, but before the onset of thermal decomposition at such a rate that it would either leave too much precursor in the film, or undergo continual "CVD-like" deposition.

However, there are other considerations that are known at the outset of measuring thermal parameters of a precursor. The TGA can give insight into slower thermal decompositions that may affect performance (through the stress test mentioned above) or the DSC might give how idea of melting point, which could aid in handling and safety. When undertaking to select a precursor, many factors can, and should be considered, and there are still many concerns about nucleation and chemisorption that need to be tested, which TGA and DSC give scant insight on. And, finally, reaction with a surface (either initially on the selected substrate, or later in the process on the growing film) must be considered: the surface is a chemical reactant like any other, and there may exist energetically accessible pathways that allow for decomposition of a chemisorbed species that are unrelated to intramolecular decomposition.

However, in the course of our research, we have found it useful to define a "Figure of Merit" for the thermal range of a precursor. It is a straight-forward method of comparing large families of precursor compounds and to select the most likely candidates:

$$\sigma = (T_D - T_V)\left(1 - \frac{\%m_r}{\%m_M}\right)$$

Here, the thermal range ($T_D - T_V$) is modified by the residual mass, $\%m_r$, and balanced for the molecular weight of the element of interest ($\%m_M$, this is almost always the metal, but generally is the payload atom that defines the film). The rationale is that the second term represents the amount of the payload atom that remains unvolatilized during the TGA experiment, and this makes the precursor less useful, and so lessening the thermal range can reflect this in a single Figure of Merit. It should be noted that there are two cases where it is possible to have a negative σ. In the first case, since T_V and T_D are measured independently of each other on separate instruments, it might be found that the onset of volatility occurs after the onset of decomposition. This is necessarily a poor ALD precursor, although it might have some utility in CVD, depending on the mechanism of decomposition. The second case involves the assumption that only the payload atom remains in the residual mass. This naturally does not have to be the case, and so a larger $\%m_r$ than the $\%m_M$ will give a negative in the second term. This also represents a poor ALD precursor, if it cannot be volatilized without a significant portion of it decomposing. Finally, it should be noted that, since there are cases where each independent term in σ can generate a negative, and this generates a false positive σ for a truly terrible precursor. This is possible using this simple Figure of Merit, and so the practitioner cannot blindly use thermal data, but rather should use this method as it is intended: to rank apparently useful precursors against each other.

5.5 Decomposition testing

Although mechanisms for decomposition and deposition will be covered in subsequent chapters, thermal testing can be used to diagnose decomposition mechanisms and can aid in the iterative design of precursor compounds.

A simple and useful method of diagnosing thermal decomposition is to dissolve the precursor in a solvent and then use a common structural diagnostic instrument (e.g., nuclear magnetic resonance, NMR) to interrogate the sample while it is being heated. The most straightforward method is to flame-seal an NMR tube and to store it in an oven at the desired temperature, periodically removing it and measuring the solution. Naturally, this relies on the experimentalist being trained to operate and interpret NMR, as well as having this instrumentation at their disposal. Thus, this method is best suited for a researcher with synthetic experience at an institution (like a University chemistry department) with the appropriate infrastructure. This method also comes with a caveat: although a dilute solution of precursor can be useful to interrogate, it is a different thermodynamic system than both the condensed phase and the gas phase, and so it may highlight thermal chemistry that is not wholly representative of the ALD process.

A good example of this is the interrogation of the surface chemistry of gold gua-nidinates, shown earlier in Chapter 4. This decomposition was found to occur by two pathways, release of a "carbodiimide," as well as beta-hydrogen abstraction (Figure 5.8).

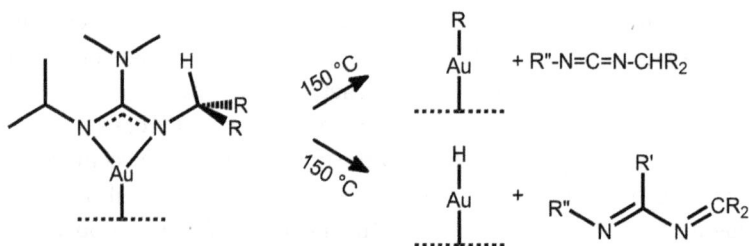

Figure 5.8: Contrasting decomposition mechanisms of a gold guanidinate species.

The top route in Figure 5.8 produces a carbodiimide species: the surface species must break an Au-N bond and allow the ligand to rotate until the -NMe$_2$ group on the bridg-ing carbon is in proximity to the gold. The gold can then make a bond to the -NMe$_2$ group and release the rest of the ligand as a carbodiimide. This is exactly the thermal decomposition seen in the solution phase, and the carbodiimide could even be iso-lated and tested to ensure that this was the by-product. It should be noted that the newly formed gold amide did not survive at 150 °C, rather it underwent a further de-composition to produce gold metal.

However, a more laborious study using a highly modified ALD reactor where the gaseous by-products could be collected and measured showed that the carbodii-mide was a minor product. The pathway was thermally viable in the ALD process, but the lower pathway (Figure 5.8) was much more prevalent, and the "oxidised guanidinate" was a much more abundant by-product. In this decomposition path-way, the organic group on the nitrogen needs to rotate until the hydrogen (drawn in the reactant in Figure 5.8) is in a proximate position. The gold center can then make a bond to this hydrogen and release the "oxidised guanidinate." Again, the gold hy-dride that results immediately reduces to produce gold metal. Both thermal path-ways exist for this surface species at 150 °C, but the lower pathway is quicker, and so becomes the predominant route of decomposition. Caution must be taken in mea-suring thermolysis.

5.6 Volatility control

Chapter 4 showed an example of melting point control, and this can be helpful in promoting volatility: it is reasonable to guess that a liquid might naturally have a higher volatility than a solid, given that the molecules in a liquid are, on average, farther apart than those in a solid. But, like all of science, this is true until it is not, and generally can be a reliable concept for structurally related molecules, but less true over all molecules. Volatility control is therefore a more difficult subject, and trial and error is almost always the preferred method of investigating how to change vapor pressure.

One reliable ploy for increasing the vapor pressure of a molecule is to reduce its "complexity": in precursors that tend to have ligands that can bridge between metal centers, reducing the ability to bridge can lessen the potency of the intermolecular interactions (as well as reduce the molecular weight) and increase volatility.

A good example of this can be found in group 13 metal species (Al, Ga, In) that have ligands containing group 15 atoms (N, P, As). These are common CVD and ALD precursors, since they can easily deposit so-called III–V semiconductor thin films. The group 15 atoms all bear a lone pair of electrons, making them particularly predisposed to bridging (Figure 5.9).

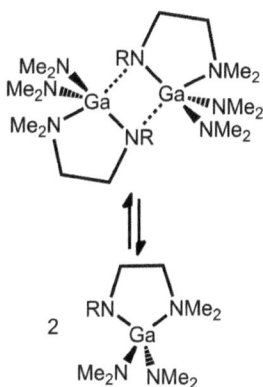

Figure 5.9: Oligomerization of a group 13-group 15 compound that hinders volatility.

This family of precursors differs by changing the alkyl group "R" in the chelate ligand RN(H)-CH$_2$CH$_2$-NMe$_2$, commonly called an ethylene diamine. The protonated end of the ethylene diamine is deprotonated to allow this compound to act as a Lewis basic ligand for gallium. When the R group is small (like methyl), the compound exists in equilibrium between the dimer (left-hand side, Figure 5.9) and two monomers (right-hand side, Figure 5.9) in solution. The small size of the R group allows for a closer contact between the lone pair on the ligand of one molecule with the metal center on another molecule. By simply changing this R group to an ethyl group, the compound exists only as monomers in solution. Simple and judicious considerations about a ligand can affect the resulting thermal properties of a pre-cursor, and this type of "design" is a complex research undertaking.

5.7 Impurity control

Another consideration in thermal testing and decomposition is the control of impurities. Although the surface chemistry of film formation and decomposition contribute many opportunities for impurities in a growing film, a common source of impurity is the atom in the ligand that is coordinated to the metal. Naturally in an ALD precursor, these bonds are strong to resist thermal decomposition during volatilization, but these bonds should be considered during thermal testing to minimize impurities down the line.

A good example of controlling impurities in this manner is the development of dihydridodimethylamidoaluminum(III) as a precursor for the semiconductor AlN (Figure 5.10).

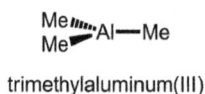

$$Me_{\textit{min}}Al{-}Me$$
$$Me$$

trimethylaluminum(III)

$$H_{\textit{min}}Al{-}NMe_2$$
$$H$$

dihydridodimethylamidoaluminum(III)

Figure 5.10: Two different aluminum precursors give different carbon impurities in resulting AlN films.

Prior to the development of this compound, AlN was made by the ALD of trimethylaluminum(III) and ammonia. This process typically leaves $10\%_{at}$–$15\%_{at}$ carbon depending on deposition temperature, due to the direct Al-C bonds found in trimethylaluminum(III). When dihydridodimethylamidoaluminum(III) is used, there are no existing AL-C bonds, and the carbon impurities drop to $< 1\%_{at}$ and cannot be detected by common film analysis methods. There are many differences between these two precursors, and it is not simply the potential impurity levels that drive precursor selection. Again, however, thoughtful precursor design, and an understanding of thermolysis, can help film deposition by avoiding conditions that can cause problems.

Chapter 6
Nucleation

When a precursor encounters a surface, there are two chemical states that need to be differentiated to discuss saturation and self-limited ALD growth: these are physisorption and chemisorption. In general, physisorption is a weak interaction of the precursor with the surface, where a chemical change to the precursor does not occur, and the process is reversible:

$$\vdash + MX_n \rightleftharpoons \vdash MX_n$$

Given the reversible nature of physisorption, the equilibrium constant (K) determines if this is sufficient to allow for saturative behavior. Generally, physisorption can be the first mechanistic step in an ALD process, but the instability makes it an unreliable reaction for saturation. Even when the equilibrium of this reaction favors the products (i.e., the surface bonded precursor, $K \gg 1$), the monolayer will start to decay as soon as the exposure of the precursor to the surface stopes. When the pulse is ended, the concentration of the reactants (i.e., free, gaseous precursor and an open surface nucleation site) changes: the pressure of gaseous precursor drops significantly, and the equilibrium will respond through the desorption of physisorbed species. However, we will see later in this chapter that physisorption of precursors, ligands, and small molecules can act on the surface to inhibit or enhance growth.

Chemisorption is generally a stronger chemical interaction with the surface, and commonly results in a chemical change in the precursor to become a surface bonded moiety. Chemisorption is characterized by being irreversible, and in general leads to saturation and self-limiting behavior as described in Chapter 2:

$$\vdash + MX_n \rightarrow \vdash MX_{n-1} + X$$

Note that this loss of a ligand does not require a loss of a surface group from the nucleation site. However, the archetypal example of chemisorption is the reaction of trimethylaluminum(III) with a hydroxyl at a silicon surface to produce methane (CH_4) in the gas phase.

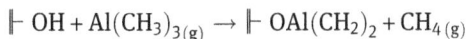

$$\vdash OH + Al(CH_3)_{3(g)} \rightarrow \vdash OAl(CH_2)_2 + CH_{4(g)}$$

Note: In this chapter I would like to acknowledge Prof. Markku Leskelä and Prof. Mikko Ritala. Their names are often mentioned together as the main, global driving force for academic ALD research, which is a true and well-earned reputation. Both of them welcomed me for a sabbatical year in the University of Helsinki, and that visit drastically altered the way I thought about ALD and my place in it. Also, I broke their Kugelrohr apparatus during my first week, and they still let me stay.

https://doi.org/10.1515/9783110712537-006

In the case of trimethylaluminum(III), and generally for nucleation that occurs with the loss of a ligand, the irreversibility is a result of the reaction conditions. Once the ligand is lost, it is carried away by evacuation or a purge gas. The system cannot revert to reactants since the ligand is no longer in the same thermodynamic system as the chemisorbed species. The surface bonded moiety therefor has a different, and typically less thermodynamically spontaneous pathway to desorption.

It is not always the case where chemisorption requires the precursor to chemically react with the surface nucleation site; it might be sufficient that the loss of a ligand (and subsequent removal of it from the thermodynamic system) renders the adsorption irreversible. Hexacarbonylmolybdenum(0) is a known precursor for binary molybdenum compounds (e.g., MoN_x, MoO_3) and coordinates to aluminum oxide at room temperature. This species loses a CO ligand, causing it to become chemisorbed, and, in the absence of CO in the gas phase, cannot readily desorb (Figure 6.1).

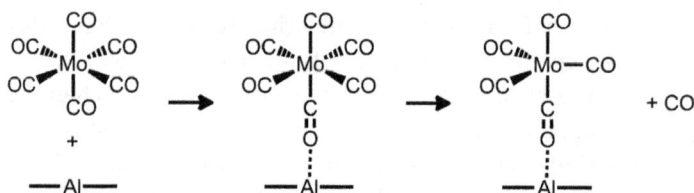

Figure 6.1: Physisorption, followed by chemisorption of hexacarbonylmolybdenum(0) at an alumina surface.

Interestingly, this precursor is found to coordinate to the metal center in the aluminum oxide surface, by bridging through a carbonyl ligand. Thus, chemisorption is not "anchored" by loss of a ligand through reaction at the nucleation site, but rather through ligand loss in a subsequent, independent step. This generates a surface moiety that is less stable in the gas phase, although it could be a neutral, independent molecule in a different thermodynamic state.

6.1 Protonated surface nucleation sites

In general, nucleation at a surface hydroxyl is the main example of nucleation that is taught in ALD. There are several reasons for this, but foremost is the presence of surface hydroxyls as nucleation defect sites on Si^0 substrates, as well as on SiO_2 and Al_2O_3 surfaces. These surfaces are important for different reasons. Si^0 is often used as a test substrate for process development due to its prevalence in microelectronic manufacturing, and reactivity at a surface ⊢Si-OH is the nucleation event that allows many ALD processes to start. Elemental silicon is quite reactive with two components of air: oxygen will oxidize the surface to SiO_2, and water will react with SiO_2 surface to produce surface hydroxyls. Finally, Al_2O_3 is an important surface since it is the growing film surface for the prototypical ALD reaction of trimethylaluminum(III) and water.

In general, any oxide surface can be considered to bear surface hydroxyls, and growth of a thin film oxide with water can be considered to have hydroxyl nucleation defects (Figure 6.2).

Figure 6.2: Surface hydroxylation by reaction with water.

This nucleation defect site is expected to also occur on metal surfaces, provided that the surfaces are oxophilic enough to react in air to form an oxide. Several other binary compounds can likewise be considered to form protonated surface nucleation sites through reaction with a small molecule precursor. For instance, a similar reaction at a nitride surface with gaseous ammonia could provide ammine nucleation sites (Figure 6.3).

Figure 6.3: Surface amination by reaction with ammonia.

An ammine-terminated surface would naturally be more complicated than that of a hydroxyl surface. The "NH" group can bear either one or two protons, thus having two reaction sites with different pK_a values.

Hydroxyl density also plays a role in nucleation at a hydroxylated surface. As was discussed in Chapter 2, the area that a precursor occupies at the surface can obscure hydroxylated nucleation sites and lower growth rate. These obscured nucleation sites can also interact with the chemisorbed surface moiety (Figure 6.4).

It was found in the case of biscyclopentyldienylbisguanidinatoTi(IV), that the surface chemistry would change depending on the hydroxyl density. It is well established that the number of hydroxyls per square nanometer on silica becomes lower at higher temperatures due to dehydration. When silica is pretreated at a lower temperature, the parent precursor retains both a guanidinate ligand and a cyclopentadienyl ligand when chemisorbed due to hydrogen-bonding of a neighboring hydroxyl group to the chelating nitrogen of the guanidinate ligand. When silica is pretreated at a higher temperature, the guanidinate ligand is not retained: the surface group keeps two ligands, but without a neighboring hydroxyl to stabilize the guanidinate, it preferentially loses these ligands and keeps the cyclopentadienyl ligands.

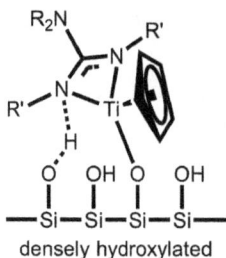

densely hydroxylated

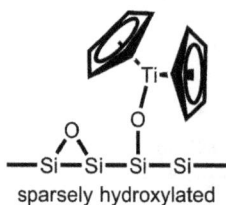

sparsely hydroxylated

Figure 6.4: The effect of surface hydroxyl density on the chemistry of chemisorption.

6.2 Metal surface nucleation sites

Metal surfaces are significantly different from protonated surfaces. On a pristine metal surface, defects in the lattice plane of the metal may provide nucleation points, as do edges (Figure 6.5).

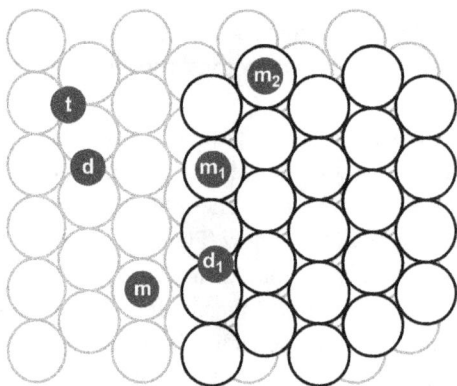

Figure 6.5: Potential nucleation sites on a metal surface.

Figure 6.5 shows a closest-packed metal surface, with each different site labeled. A precursor might coordinate at one surface atom on a face (labeled "m"), or at either a flat ("m_1") or variegated edge ("m_2"). The precursor could also bridge between a surface dimer ("d") or edge dimer ("d_1") site, or a trimer site ("t") would allow a precursor to coordinate in the three-atom "dimple" in the surface. These various sites are more chemically similar than the variety of hydroxyl groups that exist on a complicated oxide surface (like alumina, discussed in Chapter 2), but metal surfaces are also less chemically selective than hydroxyl groups. This means that a precursor might bind to a metal surface through its ligands as well as its metal center. An example of this is the interaction of bisdimethylamino-2-propoxocopper(II) with a pristine Cu(III) surface (Figure 6.6).

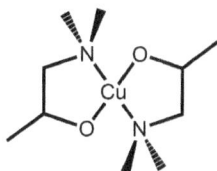

bisdimethylamino-2-propoxocopper(II)

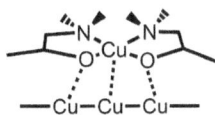

Figure 6.6: The coordination of bisdimethylamino-2-propoxocopper(II) to a copper surface.

The bisdimethylamino-2-propoxocopper(II) adsorbs to the surface a several surface sites, but in every case it coordinates through the copper center as well as the oxo moieties in the ligand. This adsorption shows the general nature of metal surface bonding and suggests that the growing copper film might get oxygen impurities through reaction with the ligand set.

6.3 Hydride surfaces

To discuss hydrogen-terminated surfaces, hydrides need to be differentiated. Inorganic chemistry considers a hydride to exist when the hydrogen is bonded to a metal center that is less electronegative than hydrogen (EN = 2.20). This definition considers most hydrogen-terminated metal surfaces as hydrides (EN = 0.70–2.16), where the electron density can be considered to reside formally on the hydrogen center, giving it a partial negative charge, and polarizing the bond. Naturally, this large range of electronegativity means that, for instance, a hydrogen terminated titanium surface (Ti-H, ΔEN = 0.66) will provide a much more active surface hydride than a hydrogen terminated molybdenum surface (Mo-H, ΔEN = 0.04). Likewise, silicon (Si-H, ΔEN = 0.30) also gives a negatively polarized surface hydride. There are notable exceptions to this: noble metals (Ru, Rh, Pd, Os, Ir, Pt, Au, EN = 2.20–2.54) and the heavy metals W (EN = 2.36) and Pb (EN = 2.33) will show a polarization favoring a positively polarized surface hydrogen, and these can be expected to act as protons. In the case where surface bond polarization favors a partial positive charge on a surface hydrogen, typical Bronsted acid/base reactivity is a reasonable guess for a chemisorption reaction.

Generally, the chemistry of a hydride surface can be considered as a reaction with H⁻ (Figure 6.7).

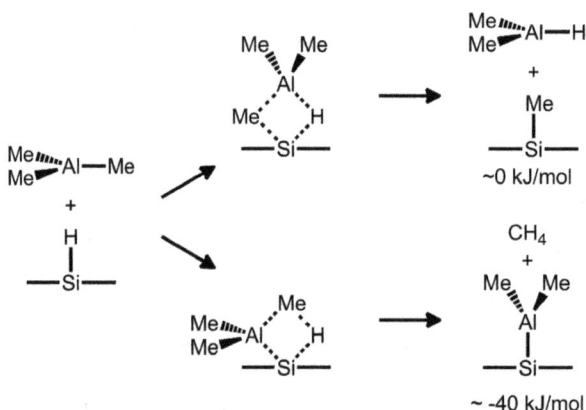

Figure 6.7: Reaction pathways for trimethylaluminum(III) with a silicon hydride surface.

This reaction typifies the useful role of hydrides for surface reactions. The upper path is a ligand exchange, where the CH_3^- and H^- exchange centers, with an overall net stabilization of zero. Although a viable equilibrium reaction, the net stabilization does not drive the reaction forward, and saturation at the surface of methyl groups will be slow. On the other hand, elimination of methane (the lower pathway) has a net stabilization of 40 kJ/mol. Here, hydrogen is a formal hydride at the surface, with the Si-H bond polarized such that the hydrogen has a hydride character. When reacted with a methyl group, the polarization of the resulting C-H bond predicts the hydrogen to have an "acid" character (i.e., H^+) due to the electronegativity of carbon (EN = 2.55). This bond is also very stable, forming a strong overlap of the participating bonding orbitals to make a strong bond. The surface bond is likewise polarized in the reverse compared to products, with the Si now bearing a partial negative charge. Thus, the intermediate that is drawn in the lower pathway represents a redox reaction, with the oxidation of the hydride and reduction of the surface silicon.

6.4 Surface blocking groups and selective area deposition

In Chapter 2, selective area deposition was presented as a kinetics argument: when two surfaces have very different nucleation rates with a given precursor, deposition will selectively occur at one surface over the other. One method of tuning a surface to exhibit two different rates of precursor chemisorption is to block the nucleation sites of part of the surface using self-assembled layers to passivate potential surface nucleation sites. If this blocking is performed with a molecule that will not allow chemisorption of the precursor, this effectively sets up a difference in chemisorption kinetics that favors selective area deposition.

The most well-studied surface blocking chemistry is the use of long chain thiols, although other blocking agents can be used. In most examples, two intimately patterned surfaces (e.g., a metal and an oxide) might both participate in ALD, albeit with different surface chemistries and potentially different GPCs. A thiol can be used as a precursor to selectively react with the metal surface, while not nucleating on the oxide surface. The surface blocking precursor would need to exhibit selectivity between these two surfaces. A good example of this is dodecanethiol blocking a copper metal surface while leaving a SiO_2 surface to react (Figure 6.8).

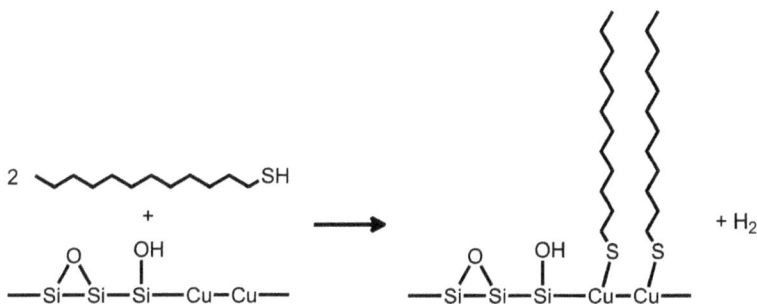

Figure 6.8: Selective reaction of dodecanethiol with Cu^0 over SiO_2.

This selectivity can be considered from an electronegativity analysis. The thiol head group is inclined to maintain a negative charge, and preferentially give up its hydrogen as a proton due to the higher electronegativity of sulphur (EN = 2.58) compared to hydrogen (EN = 2.20). The copper surface can accommodate a negatively charge sulphur head, having a lower electronegativity (EN = 1.90), but the oxygen at the silicon surface cannot (EN = 3.04). Since the long, alkyl tail of the dodecanethiol terminates in a CH_3 group, there is no chemical ability of this surface moiety to allow chemisorption of most precursors, effectively blocking the surface from deposition. The study of self-assembled monolayers, particularly using thiols and a metal surface, is a mature field of research. Interestingly, when the thiol coordinates to the metal surface, the fate of the hydrogen atoms is not entirely clear. Dihydrogen can be detected as a by-product, but never in a concentration that suggests that this is the sole fate of the hydrogen. Other suggestions include having the hydrogen still surface bonded as hydrides, or having it dissolve into the metal, either as H_2 or $H \cdot$ radical. However, this self-assembled monolayer blocking is an effective strategy to improve selective area deposition.

Chapter 7
ALD processes

There are many different ALD processes with unique surface chemistries and precursors. In this chapter, we will examine some unusual processes in depth to bring a better understanding of the commonalities, as well as the uniqueness of these processes. This chapter will highlight several fundamental mechanistic steps from inorganic chemistry and highlight how they apply to surface chemistry (Table 7.1).

Table 7.1: Fundamental chemical reactions in inorganic chemistry.

Type	Reaction	Effect on Oxidation State
Association	$ML_n + L \rightarrow ML_{n+1}$	None
Dissociation	$ML \rightarrow ML_{n-1} + L$	None
Comproportionation	$M^{(n-x)+} + M^{(n+x)+} \rightarrow M^{n+}$	Decreases/Increases
Disproportionation	$M^{n+} \rightarrow M^{(n-x)+} + M^{(n+x)+}$	Increases/Deceases
Oxidative Addition	$ML_n + X_2 \; ML_nX_2$	Increases
Reductive Elimination	$ML_nX_2 \rightarrow ML_n + X_2$	Decreases
Migratory Insertion	$ML_nX \rightarrow ML_{n-1}(X-L)$	None
Beta Elimination	$ML_{n-1}(X-L) \rightarrow ML_nX$	None

These fundamental mechanistic steps come in pairs that are the opposite of each other. The examples that follow highlight these fundamental reactions and explain them fully, in the context of ALD surface chemistry.

7.1 Alumina deposition

7.1.1 Trimethylaluminum(III) and water

The standard example of an ALD of Al_2O_3 is trimethylaluminum(III) and water: the surface chemistry occurs through straightforward Bronsted acid/base chemistry, and the stoichiometry works out in a straightforward manner:

Note: In this chapter, I would like to acknowledge Prof. Roy Gordon, my former post-doctoral supervisor. He got into the field of ALD while I was in his group and dragged me along with him. His intelligence, insight, and demeanor have had a tremendous influence on me, and I was lucky to work in his group.

https://doi.org/10.1515/9783110712537-007

$$3 \vdash OH + 2AlMe_3 \rightarrow \vdash OAlMe_2 + (\vdash O)_2 AlMe + 3\ CH_4$$

$$\vdash OAlMe_2 + (\vdash O)_2 AlMe + 3\ H_2O \rightarrow 3 \vdash OH + 3\ CH_4$$

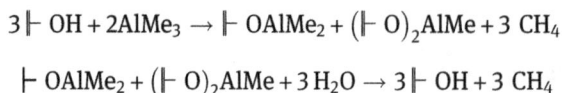

Even the complication that trimethylaluminum(III) reacts at either one or two $\vdash$OH groups can be simply accommodated in a straightforward chemical reaction with only a minor understanding of surface chemistry. Hydroxyl surfaces are also known to dehydrate with increasing temperatures, with two adjacent hydroxyls forming a bridged oxo and water:

$$2 \vdash OH \rightarrow \vdash O + H_2O$$

This accounts for the drop-off in growth per cycle that occurs when alumina is deposited at higher temperatures from this process.

7.1.2 Trimethylaluminum(III) and ozone

The surface chemistry is not as straightforward for the ALD of Al_2O_3 from trimethylaluminum(III) and ozone. Ozone can itself react at a surface, or it delivers an elemental oxygen:

$$O_3 \rightarrow O_2 + O$$

The most cited surface chemistry is "incomplete combustion" of the surface bonded methyl groups to produce CO, CO_2, and H_2O:

$$\vdash OAl(CH_3)_2 + 6\ O \rightarrow \vdash OAl(OH)_2 + CO + CO_2 + 2\ H_2O$$

The preceding chemical equation is balanced but does not represent reality: the combustion of the surface is a complicated combination of reactivity that changes with temperature, exposure, system pressure, and abundance of O_3 and elemental oxygen in the gas phase. A closer look shows individual reaction pathways, but it is worth noting that these may or may not contribute in a meaningful way depending on the reaction conditions.

The nucleation of trimethylaluminum at a hydroxylated surface happens following the typical Bronsted chemistry (Figure 7.1a). This is notable, since combustion of the surface ligands by ozone does not necessarily result in hydroxyl groups.

Ozone is known to react with trimethylaluminum(III) to produce a methoxy group (Figure 7.1b). The reaction pathway can be imagined going through ozone, or a released oxygen atom, coordinating to the empty π orbital on the aluminum center, followed by the migration of the methyl group to the oxygen. This appears as an "insertion" of the oxygen into the Al-C bond; it is formally a 1,1-migratory insertion (see Table 7.1), where the metal center and the ligand both shift to bond to the

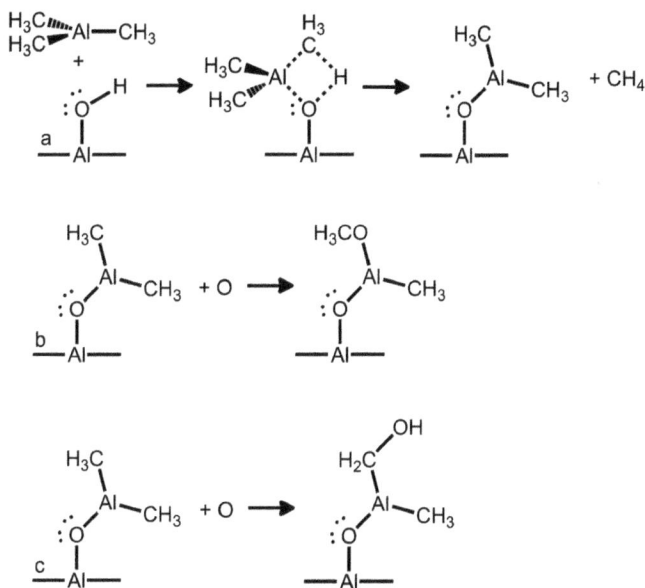

Figure 7.1: Initial reactions in the trimethylaluminum(III) + ozone ALD surface reaction.

same center on an added ligand. Here, the nomenclature is not strictly necessary, since there is only one atom in the added group. Migratory insertion is a common mechanistic step in organometallic chemistry, where a ligand at a metal center shifts onto another existing ligand (Figure 7.2).

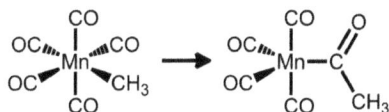

Figure 7.2: Idealized 1,1-migratory insertion at a metal center.

A second identified path is the same type of insertion of the oxygen into a C-H bond (Figure 1c). As more ozone is added to the surface-bonded aluminum, each ligand, and the surface bond itself (if it is not already oxo containing), can undergo a similar migration step to form many alkoxy and hydroxy groups at the surface aluminum. Further evidence by IR has shown that surface formate and bicarbonate groups can be detected (Figure 7.3).

In the presence of a surface hydroxyl, two neighboring methoxy groups will form a formate moiety when one methyl group eliminates as methane (Figure 7.3a). Subsequent reaction of the surface formate with an oxygen atom can produce a bicarbonate (Figure 7.3b). Both groups can subsequently produce a surface hydroxyl by elimination of CO (from formate) or CO_2 (from bicarbonate):

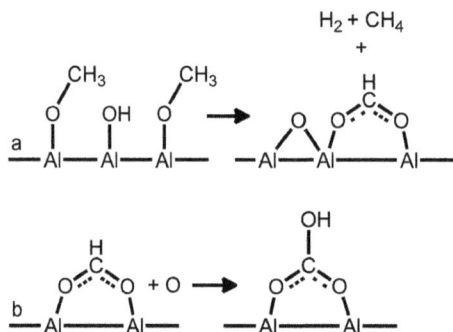

Figure 7.3: Formation of formate and bicarbonate groups in the reaction of surface methoxy groups with elemental oxygen.

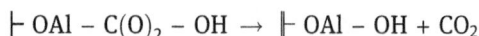

$$\vdash OAl - C(O)OH \rightarrow \vdash OAl - OH + CO$$

$$\vdash OAl - C(O)_2 - OH \rightarrow \vdash OAl - OH + CO_2$$

Once hydroxyl groups are produced, there are two subsequent reactions that can occur: a surface methyl group can be protonated to produce methane, or dehydration of two hydroxyl groups can produce water. This regeneration of surface hydroxyls allows the process to continue in a cycle-by-cycle manner insomuch as TMA can now again undergo classical a Bronsted acid/base reaction to chemisorb. This explains why there is a similar temperature dependence on growth per cycle as is seen in the TMA/water process.

A less likely pathway involves the shift of the hydroxyl group from the $\vdash CH_2OH$ group to an open surface aluminum:

$$\vdash Al + \vdash AlCH_2OH \rightarrow \vdash AlOH + \vdash AlCH_2$$

This generates a stable surface hydroxyl, but also an unstable CH_2 "carbene," which will quickly combine with another CH_2 group to make ethene:

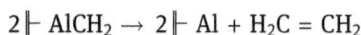

$$2 \vdash AlCH_2 \rightarrow 2 \vdash Al + H_2C = CH_2$$

Ethene is detected as a gas phase by-product early in the ozone pulse of Al_2O_3 deposition, and it is suspected that this process becomes less viable as surface hydroxyls become more plentiful.

The deposition of alumina using TMA thus shows two extremes of reactivity: the straightforward reaction with water, and the multi-path surface reactivity with ozone. It should be noted that since ozone is generally thought to react by generating a radical $O \cdot$, the chemistry described here for ozone can also be considered to occur using an oxygen plasma.

7.2 Titanium nitride deposition

A key issue of the deposition of titanium(III) nitride deposition is the reduction/oxidation chemistry that occurs during the process. Precursors with a Ti^{3+} center are rare, due to the relative thermodynamic instability of this center compared to Ti^{4+}. The higher oxidation state precursors are favored in most depositions of titanium-containing films because Ti^{4+} precursors have better thermal stability in the bubbler than Ti^{3+} precursors. Indeed, deposition of TiO_2 from $TiCl_4$ (sometimes called "Tickle") and water follows ideal Bronsted acid/base chemistry:

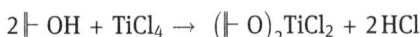

$$2\models OH + TiCl_4 \rightarrow (\models O)_2 TiCl_2 + 2\,HCl$$

$$(\models O)_2 TiCl_2 + 2\,H_2O \rightarrow 2\models OH + 2\,HCl$$

In the case of the oxide deposition, the oxidation state of the titanium in the precursor is preserved. This is different in the deposition of the nitride. In the case of TiN deposition from $TiCl_4$ and NH_3, the overall reaction is commonly written as:

$$TiCl_4 + 2\,NH_3 \rightarrow TiN + 4\,HCl + H_2 + 1/2\,N_2$$

Thermodynamic analysis of the most likely products has shown this to be the most thermodynamically spontaneous reaction. Naturally, N_2 is difficult to quantify in an ALD process, but dihydrogen is not commonly found with standard deposition conditions. There has been some speculation that the initial deposition indeed deposits Ti_3N_4 preserving the metal's oxidation state, and these only become reduced as subsequent ALD cycles, or exposure to atmosphere, proceed.

One surface process that has been proposed for this reaction is reductive elimination (see Table 7.1). This is a common mechanistic step in organometallic chemistry, where two ligands on a metal combine to form a stable molecule, and the metal center gets reduced by two electrons (Figure 7.4).

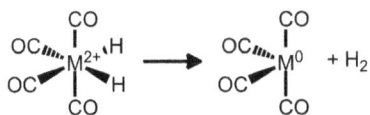

Figure 7.4: Idealized reductive elimination from a metal center.

Reductive elimination could account for the loss of dinitrogen and dihydrogen during the transformation of Ti_3N_4 to TiN (Figure 7.5a). There have also been proposals that a chloramine species could be formed to correct the stoichiometry (Figure 7.5b).

In the case of dinitrogen formation, the loss of dinitrogen reduces the Ti^{+4} to Ti^{3+} while oxidizing the nitrogen from N^{3-} to N^0 (in N_2):

$$2\models Ti_3N_4 \rightarrow 2\models Ti_3N_3 + N_2$$

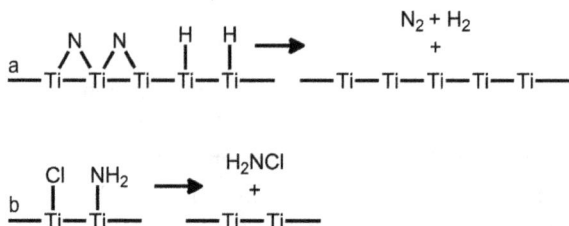

Figure 7.5: Surface mediated reductive eliminations to form TiN.

A strong driving force for this reaction is the formation of very stable dinitrogen. In the case of dihydrogen, a surface amine could transfer a hydrogen to the surface where it formally adopts a negative charge, becoming a hydride:

$$\vdash Ti - NH_2 + \vdash Ti \rightarrow \vdash Ti = NH + \vdash Ti - H$$

The reduction of titanium then comes coupled to the oxidation of H^- to H^0 (in H_2).

The case of forming a chloramine is less obvious. From the surface chemistry, both the NH_2^- and Cl^- moieties can be considered to be negative, so when the N-Cl bond is formed, two electrons need to be left behind at the surface. From a strictly electronegativity point of view, the amide fragment in chloramine would be assigned to be partially positive, and the chloride as partially negative.

Stronger support for a reductive elimination pathway comes from the deposition of TiN from Ti(NMeEt)$_4$ and NH$_3$. After the standard nucleation of the tetraaminotitanium(IV) at a growing TiN surface, there was evidence of both a chemisorbed organic imine, and the conjugate acid of the amine ligand in the gas phase (Figure 7.6).

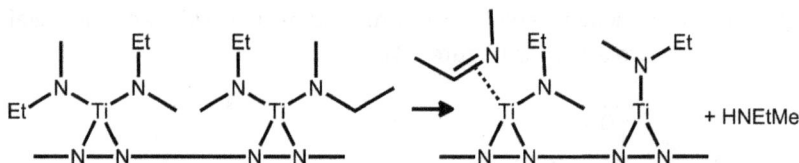

Figure 7.6: Reductive elimination of surface amines in TiN deposition.

Reductive elimination is one straightforward surface mechanism for the change in an oxidation state during an ALD process.

Simple reduction is a contrast to reductive elimination. In this case, the precursor is reduced by an incoming second precursor, commonly by a ligand exchange. A good example of this is the deposition of Cu^0 from chlorocopper(I) and zinc metal:

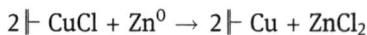

$$2 \vdash CuCl + Zn^0 \rightarrow 2 \vdash Cu + ZnCl_2$$

Chlorocopper(I) can saturated a copper surface, effectively making it chloride terminated. Zinc metal melts at 420 °C and has sufficient vapor pressure to be used as a vapor phase precursor at 435 °C. Here, the zinc reacts directly with the chloride layer, taking two chlorides while being oxidized for two electrons. The surface is subsequently reduced.

7.3 Copper metal deposition

Besides simple reduction of chlorocopper(I), Cu^0 can also be deposited by reductive elimination. An ALD precursor cannot carry two different ligands that would undergo reductive elimination in an intramolecular or even intermolecular fashion. This would allow the precursor to decompose to the metal continually, and lead to CVD deposition. In the case of the ALD of Cu^0, reductive elimination can be made to occur after ligand exchange. When bis(dimethylaminoethoxo)copper(II) is used in copper metal deposition, the first step of the reaction is for the precursor to lose a ligand to the surface (Figure 7.7).

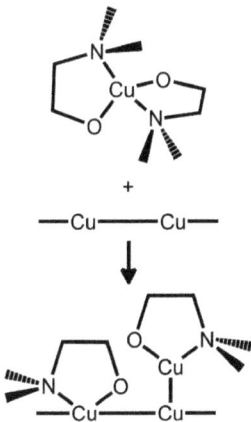

Figure 7.7: Chemisorption of bis(dimethylaminoethoxo)copper(II) at a copper surface.

When the ligand migrates to the surface, a copper atom on the surface (Cu^0) and the copper "adatom" (from the precursor, Cu^{2+}) undergo comproportionation (see Table 7.1), rendering both metal centers as formal monocations:

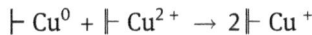

$$\vdash Cu^0 + \vdash Cu^{2+} \rightarrow 2\vdash Cu^+$$

These surface species can then undergo ligand exchange with diethylzinc(II), to exchange the surface dimethylaminoethoxides (dmae) with ethyl groups. These ethyl groups subsequently undergo reductive elimination to produce butane and leave behind Cu^0:

$$2 \vdash Cu(dmae) + Et_2Zn \rightarrow 2 \vdash CuEt + (dmae)_2Zn$$

$$2 \vdash CuEt \rightarrow 2 \vdash Cu^0 + butane$$

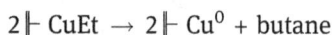

The surface ⊢Cu(dmae) groups can also react with a simpler reducing agent, dihydrogen, to produce the conjugate acid of the ligand (Hdmae) and a surface copper hydride, which can then undergo reductive elimination to produce copper metal:

$$\vdash Cu(dmae) + H_2 \rightarrow \; \vdash CuH + Hdmae$$

$$2 \vdash CuH \rightarrow 2 \vdash Cu^0 + H_2$$

Finally, if the surface is hot enough, the ⊢Cu(dmae) can undergo beta-hydrogen elimination (see Table 7.1) to produce a surface hydride, which again reductively eliminates. Beta-hydrogen elimination is a common decomposition pathway where a hydrogen atom two bonds away from a metal center can be abstracted by a metal center (Figure 7.8).

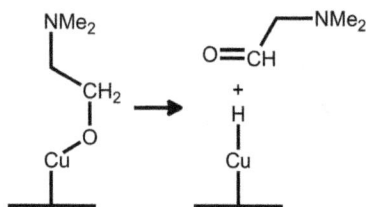

Figure 7.8: Beta-hydrogen elimination from dmae to produce a surface hydride.

This reaction is not self-limiting, and so the compound bis(dimethylaminoethoxo) copper(II) can exhibit self-limiting ALD at lower temperatures but undergoes CVD at higher temperatures. This wide variety of surface reactivity available to Cu(dmae)$_2$ makes it a very versatile process for many copper metal deposition applications, but it is limited by this CVD process to low-temperature applications only. Notably, beta-hydrogen elimination is often a low-temperature decomposition route, and is shown already in Chapter 5 (Figure 7.8).

It is worth mentioning here that disproportionation (see Table 7.1) is a common reaction with Cu(I) precursors, although it has been found that using nitrogen-based ligands suppresses this disproportionation:

$$2\,CuL \rightarrow Cu^0 + CuL_2$$

This reactivity can be detrimental to ALD of Cu^0 since disproportionation would be a continual process. Indeed, betadiketonate copper(I) compounds are excellent CVD precursors for copper metal due to disproportionation.

7.4 Gold metal deposition

Gold metal can be deposited from trimethyltrimethylphenylgold(III) and a selection of common second precursors: oxygen plasma (with water as a necessary component), ozone, and hydrogen plasma. The first step of this reaction is dissociation of the precursor on a gold surface:

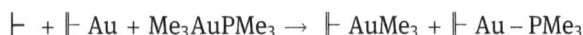

$$\vdash + \vdash Au + Me_3AuPMe_3 \rightarrow \vdash AuMe_3 + \vdash Au - PMe_3$$

In organometallic mechanistic chemistry, association and dissociation are two of the most fundamental synthetic steps (see Table 7.1). Ligand exchange reactions can be characterized as being fully associative, fully dissociative, or intermediate between the two (Figure 7.9).

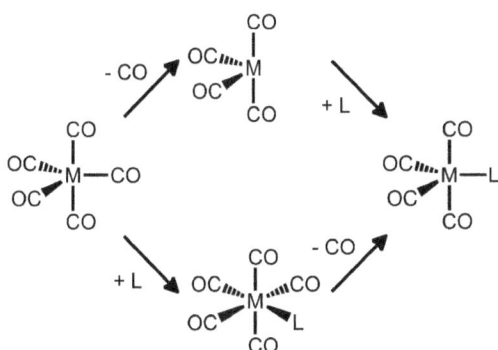

Figure 7.9: Idealized associative and dissociative ligand exchange from a metal center.

Notably, all ALD processes have an inherent association step: the physisorption of a precursor, or even a precursor fragment, at a surface is an association step. In general, dissociation that occurs at a surface can have deleterious effects. If the ligand that is transferred to the surface is adsorbed strongly, it can both passivate the surface, making chemisorption of the precursor more difficult, and it can also add impurities to the growing film if it decomposes.

Strong adsorption of the phosphine occurs during the deposition of gold metal from Me_3AuPMe_3. In an ALD process using just oxygen plasma, gold metal is formed, and the ligand system is combusted to CO_2, CO, and water (as discussed above for ozone). Without additional water, there is also P_2O_5 left in the film as an impurity from phosphine combustion, which is removed when reacted with a pulse of water:

$$2 \vdash P_2O_5 + 3 H_2O \rightarrow \vdash + 2 H_3PO_4$$

Indeed, the dissociation of the ligand system is quite noticeable when using this precursor for gold deposition. Both carbon and phosphorous impurities are found when using only oxygen plasma as a precursor. These impurities are diminished significantly when hydrogen plasma is used. By careful monitoring during the deposition of Au^0 with H_2 plasma as the second precursor, it was found that both trimethylphosophine and methyl ligands dissociate from the trimethyltrimethylphosphinegold(III) precursor onto the growing gold surface (Figure 7.10).

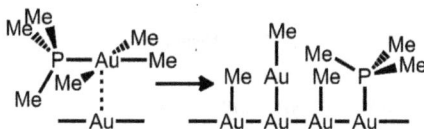

Figure 7.10: Ligand dissociation from trimethyltrimethylphosphinegold(III) at a gold surface.

The hydrogen plasma removes the methyl groups as methane and combusts the phosphine into methane and lower-order phosphines (i.e., Me_2PH, $MePH_2$, PH_3).

7.5 Tin oxide deposition

Tin oxide can typically be deposited with the Sn center in either the + 2 or the + 4 oxidation center, resulting in SnO or SnO_2, respectively. Selecting a process that can deposit "phase pure" material can be difficult. If the target film is SnO_2, an oxidizing environment is important, and so using a precursor like peroxide can help. Peroxides are chemical compounds with an existing O-O single bond, giving the oxygen a formal oxidation state of + 1. The deposition of SnO_2 from an N-heterocyclic silylene precursor (N^2,N^3-di-*tert*-butyl-butane-2,3-diamido-tin(II), the "stannylene") and hydrogen peroxide goes through an oxidative addition mechanism (see Table 7.1). Oxidative addition is the opposite mechanistic step to reductive elimination (Figure 7.11).

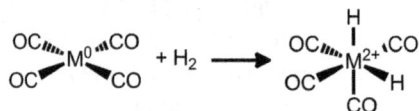

Figure 7.11: Idealized oxidative addition at a metal center.

In the deposition of SnO_2, the first step is coordination of the stannylene to two surface hydroxyl groups, and subsequent loss of the doubly-protonated conjugate acid of the ligand (Figure 7.12a).

The surface bonded tin(II) ion reacts with one equivalent of hydrogen peroxide via oxidative addition to form a gem-alkoxy surface group (i.e., a surface atom with two nucleation sites). This oxidizes the tin center from Sn^{2+} to Sn^{4+}. The surface gem-alkoxide can then further react- which to form bridging oxo groups to other tin centers, and preserve surface nucleation site, allowing for the cycle to be repeated by standard Bronsted acid/base chemistry.

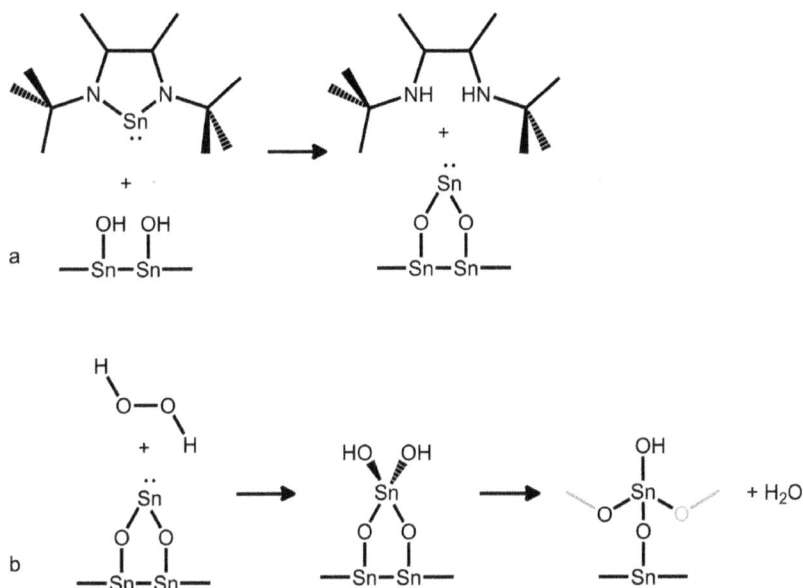

Figure 7.12: Deposition of SnO$_2$ from a stannylene and hydrogen peroxide.

This selection of ALD surface mechanisms is not at all complete but was selected to highlight and identify the classic organometallic reactions that can occur either during ALD or during surface decomposition.

Chapter 8
MLD processes

Molecular layer deposition (MLD) is a type of ALD where large organic layers can be deposited, depending on choice of precursors. The fundamental principles of the process come from ALD, and MLD processes undergo self-limiting, saturative growth that is dependent on exposure of a surface to two or more precursors, in sequence. Like ALD, it is typically based on gas-to-surface reactions, and growth is measured in Å as a GPC, rather than in Å/minute as a growth rate. Notably, the inclusion of large organic molecular precursors in thin film growth means that the GPCs of MLD processes are widely varied and typically much larger than in ALD processes. This makes MLD an interesting subfield for most ALD scientists to explore: the fundamental training, and in many cases, the equipment, translates between ALD and MLD easily.

In this chapter, typical Bronsted acid/base chemistry will be shown for the deposition of inorganic/organic hybrid films, and more involved surface chemistry will be highlighted using unique organic chemistry related to the organic precursor(s) used in MLD. This chemistry can be quite different from traditional ALD, and the variety of chemistry can encompass all organic chemistry transformations. This demonstrates the significant promise of MLD which is presently being explored by many research groups worldwide.

8.1 Alucone deposition

On major difference with MLD is the conceptualization of what can be a precursor. For example, a material based on Al_2O_3 can be deposited from TMA and ethylene glycol (i.e., 1,2-ethanediol, Figure 8.1).

This chemistry is naturally modeled after the reaction of TMA and water. The reactivity of the TMA dose is very similar to that seen in the deposition of alumina, which is a Bronsted acid/base exchange at the surface where the TMA loses either one or two methyl groups as methane, leaving a methyl terminated surface. In the second pulse, ethylene diol reacts to liberate the remaining methyl groups and regenerate a hydroxyl surface (Figure 8.1). The ethylene glycol is a "diol," where is has two hydrogens (on separate hydroxyl groups) to participate in Bronsted acid/base chemistry. Overall, this process can have a growth-per-cycle of 4.0 Å, which is quite high compared to typical oxide depositions by ALD (i.e., 1 Å). These types of MLD films are referred to as

Note: In this chapter, I would like to acknowledge Prof. Steven George. His work in this field has been ground-breaking, but my gratitude is for his friendly and approachable nature. Indeed, while writing this chapter, I asked him a question about the etymology of "metalcone": it turns out he made that word up.

https://doi.org/10.1515/9783110712537-008

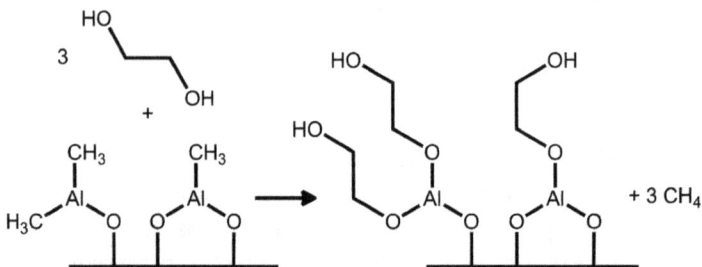

Figure 8.1: Chemisorption of ethylene glycol to a monolayer of TMA fragments.

"metalocones," a name coined by the George research group (University of Colorado) to correspond to the trivial name for the aluminum-containing film alucone. MLD was proposed in 1991 for the deposition of imide polymer films and became a focus within the ALD community starting in about 2008 with the deposition of alucone. Many other metals (e.g., Ti, V, Mn, Zn, Zr, Sn, Hf) have also been used in similar depositions of various metalcones.

8.1.1 Bifunctional monomers

There is an inherent problem in bifunctional monomers for MLD that does not exist with water: when the backbone of the molecule gets too long and exhibits flexibility, the molecule contorts to react twice with the surface, quenching their functional groups and preventing continued growth (Figure 8.2).

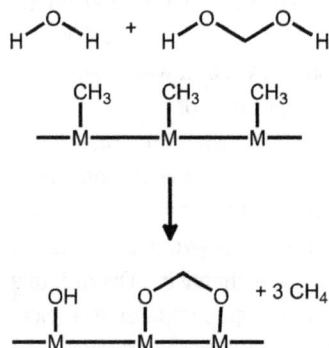

Figure 8.2: Longer bifunctional precursors may be quenched at the surface.

There are two fundamental solutions to this: to alter the organic precursor so that the functional group is not the same on both ends of the molecule, and so it cannot undergo two reactions with the same nucleation defect at the surface, or to stiffen the backbone of the organic molecule so that it cannot bend to react with the surface with both ends. The following examples highlight these solutions.

Given the complexity available in the organic precursors used in MLD, there are many possible deposition mechanisms that exploit organometallic chemistry. Another way to deposit alucone films is to exploit the reactivity of an epoxy ring, like that found in glycidol (Figure 8.3).

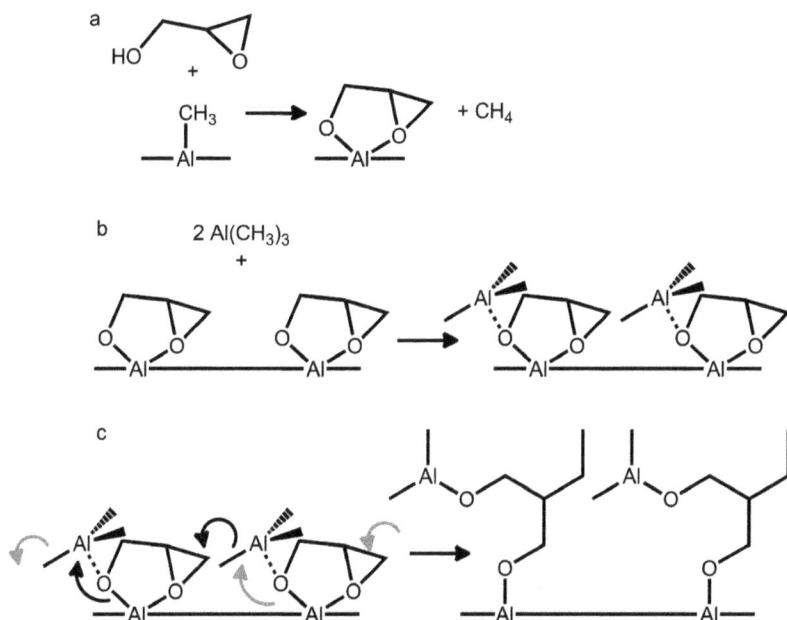

Figure 8.3: Alucone deposition using TMA and glycidol.

The TMA in this case undergoes a Lewis acid reaction (i.e., accepts a pair of electrons) from the oxygen of the ligand (Figure 8.3a). During this same pulse, the chemisorbed TMA transfers a methyl to the exposed carbon of the epoxide ring, breaking the O-C bond and forming a new ethyl branch. This then also causes the surface aluminum to make a covalent bond with the epoxide oxygen, allowing the other oxygen to form a covalent bond with the newly adsorbed aluminum (Figure 8.2b). Notice that this reaction has a cascade effect, where each TMA loses a methyl to the gycidolide group adjacent to it. It is notable that this all occurs during the TMA pulse. The glycidol pulse simply has the glycidol undergoing standard Bronsted acid/base chemistry with the remaining aluminum bonded methyl groups to exchange the surface groups and form methane (this is not shown in Figure 8.3).

The methyl group attacks the "apical" carbon of the epoxide ring (i.e., the carbon on the free corner of the triangle), and so this is a "C2" epoxy ring opening. The methyl group from the aluminum is a strong nucleophile and attacks the C2 in a classic organic "S$_N$2" reaction. The driving force for S$_N$2 reactions is the development of a stable leaving group, and here the leaving group is the oxygen already coordinatively bonded to the surface aluminum. When the epoxide ring is opened, this transforms into a strong covalent bond to aluminum, which reacts as a classic oxophile in this example. Indeed, both oxygen groups in the glycidol ligand become covalently bonded to (different) aluminum centers, giving a very strong overall thermodynamic driving force for this reaction.

8.2 Tetracyanoethylene deposition

Due to the organic-inorganic nature of MLD, some target films that were previously unobtainable are now accessible. One class of films that MLD accesses is tetracyanoethylenes, which can form organic based magnets with transition metals like vanadium and cobalt. The MLD processes for M(tcne)$_x$, where M can be Co or V, and "tcne" is tetracyanoethylene, uses metal carbonyls as their starting material. Chapter 6 (Figure 8.1) highlighted the coordination of carbonyl species: these proceed through loss of CO to deposit metal in a CVD-like reaction, and generally only undergo oxidation of the metal center when there is a high hydroxyl density at the surface. In general, the use of carbonyl precursors requires pre-treatment of the substrate for deposition to ensure that the metal centers stay in the zero oxidations state. Since these chemisorbed metal carbonyl species form metal nanoparticles, they must ultimately lose the carbonyl that anchors them to the surface (Figure 8.4a).

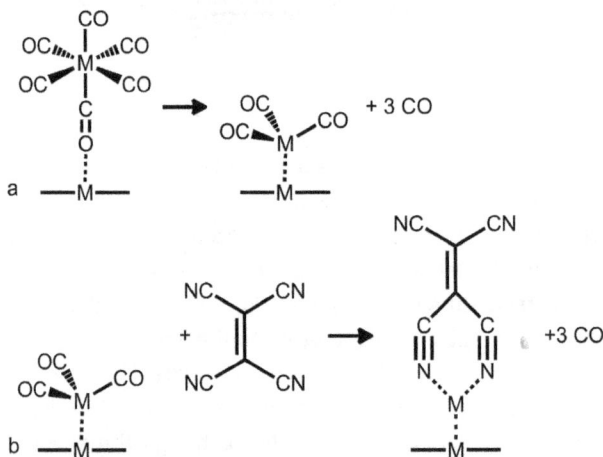

Figure 8.4: Nucleation of metal carbonyls and their reaction with tetracyanoethylene at a metal surface.

After nucleation, the tetracyanoethylene is introduced, and can coordinate, displacing the rest of the carbonyls (Figure 8.4b). Once nucleated in this manner, a following pulse of a metal carbonyl will lose CO to coordinate to the open end of the tetracyanoethylene, continuing the MLD process.

Both CO and R-CN ligands will undergo back-bonding (discussed in Chapter 3). Although CO is a stronger bonded ligand to the metal center, breaking the symmetry of the precursor (from a highly symmetric octahedron in $M(CO)_6$) weakens the M-CO bond, and lets the tetracyanoethylene coordinate. From there, the chelate effect (also discussed in Chapter 3) becomes a driving force for the coordination of tetracyanoethylene, releasing more of the CO ligands. In traditional solution chemistry, an equilibrium would be established between these two ligands, but loss of the CO to the gas phase, and ultimately out of the ALD reactor, pushes this reaction to completion by Le Chatelier's principle.

Here, the double bond between the carbons that bear the cyano groups ensures that there is no flexibility in the precursor. The precursor cannot quench both of its coordinating ends since they both cannot contact the surface at the same time.

8.3 Alq₃ deposition

Tris(8-hydroxyquinolinato)aluminium(III) is colloquially known as Alq₃, and is a common component in organic light emitting diodes (OLEDs). It can be made by CVD in a co-condensation reaction, where the hexa-coordinate aluminum species (Figure 8.5) is formed at or near a surface, and the material condenses into a thin film.

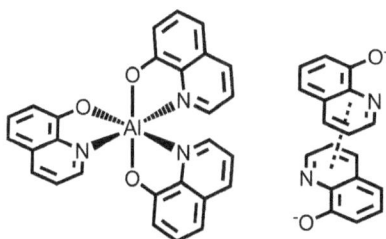

Figure 8.5: The structure of Alq₃ and q-q pi interactions.

Although the hydroxyquinolinate ligand is bifunctional, and so could be used as a precursor for MLD, this would require some of the hydroxyquinolinate ligands to bridge between metal centers, and the structure of Alq₃ is known to be a discreet molecule that crystallizes with some "pi-stacking," where the aromatic groups of the ligand overlap with neighboring ligands. The process also appears to undergo etching when too much TMA is introduced, possibly through ligand exchange:

$$\vdash Alq_3 + Al(CH_3)_3 \rightarrow \vdash + q_2AlCH_3 + qAl(CH_3)_2$$

All of this suggests that the TMA pulse undergoes ligand exchange with the Alq$_3$ surface, and the newly exchanged species adsorbs through pi-stacking to the surface (Figure 8.6a).

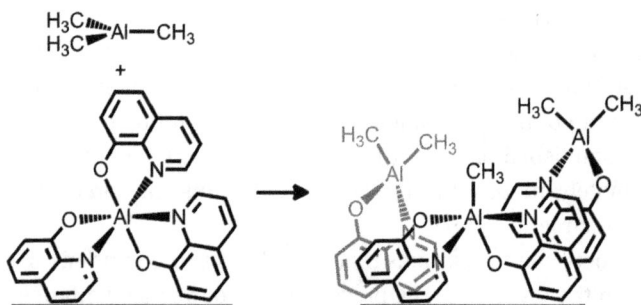

Figure 8.6: The ligand exchange of Alq$_3$ and Al(CH$_3$)$_3$ during MLD.

Pi stacking occurs when two aryl systems share the pi orbitals above and below the aromatic ring (Figure 8.5). This interaction is only on the order of 10 kJ/mol, making it a weak bond. Hydroxyquinoline pi stacks such that its nitrogen-containing rings stack "head-to-tail" and this can hold it in the solid phase (mp = 237 °C). This pi-stacking adsorption would be a weaker, physisorbed interaction compared to most ALD and MLD chemisorption interactions: this process is known to show a significant drop in growth per cycle as temperature increases, supporting the concept that the nucleation of the surface qAl(CH$_3$)$_2$ group is through a weaker and more reversible physisorption.

The subsequent pulse of hydroxyquinolinate then reforms Alq$_3$ at the surface. This step is not shown in Figure 8.5 since it would be quite congested to draw correctly. It is notable that the Alq$_3$ molecules would reorganize during deposition into their preferred phase (the α-phase), and this is a necessary condition for this film to undergo the photoluminescence that makes it a valuable material for OLEDs.

This deposition highlights a major difference with MLD: when larger organic groups are introduced, new and varied methods of chemical interaction and bonding at the surface are introduced as well. Although this gets away from more straightforward ALD principles, it also opens many more materials and surface mechanisms to study.

8.4 Kevlar deposition

Kevlar is a fabric made of the nylon-derived polymer poly(p-phenylene terephthalamide) (PPTA). Part of the strength of Kevlar comes from the fact that the polymer crystallizes when drawn into fibers, giving it a high toughness. This polymer is made through the copolymerization of terephthaloyl chloride and para-phenylenediamine (Figure 8.7).

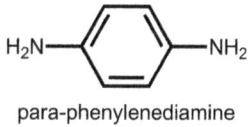

para-phenylenediamine

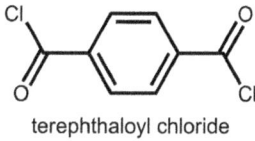

terephthaloyl chloride   **Figure 8.7:** The monomers needed to fabricate Kevlar.

Generally, this polymerization occurs as the condensation between the acid chloride and the diamine. These alternate in the polymer chain, and it becomes obvious how this might be converted into an MLD process. Notably, both precursors in this example are organic, and this example demonstrates how thin polymeric films can be deposited by MLD. Indeed, polymer deposition by MLD is becoming a more prevalent topic, and it holds significant promise for new materials and surface interfaces.

On a growing polymer surface, the terephthaloyl chloride undergoes a Bronsted acid/base chemisorption with a surface amine group (Figure 8.8a).

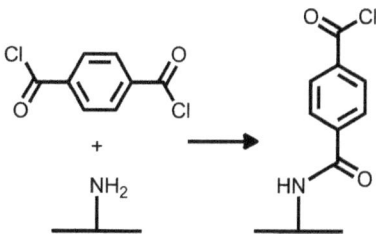

Figure 8.8: The MLD process for "Kevlar" at a surface.

This eliminates hydrogen chloride and leaves an open acid chloride group for nucleation. In the case of poly(p-phenylene terephthalamide), both monomers have a phenyl group in their body, forcing a strictly linear geometry in each precursor. This is important for polymer chain growth, since the incoming precursor cannot easily quench on the surface as depicted in Figure 8.2.

The second precursor undergoes the same Bronsted chemisorption reaction, again eliminating HCl, and regenerating the $\vdash$NH$_2$. This is very straightforward and traditional ALD surface chemistry but demonstrates the ease with which polymerization processes might be transform into MLD processes if the individual monomers exhibit the necessary precursor characteristics of low onset of volatility (T_V) and high onset of decomposition (T_D).

8.5 PEDOT deposition

Poly(3,4-ethylenedioxythiophene) (PEDOT) is a conductive polymer that is used in flexible electronics and prized for its low cost and excellent performance. Unlike the example of Kevlar, this polymer is constructed only from one monomer, ethylene-dioxythiophene (EDOT, Figure 8.9).

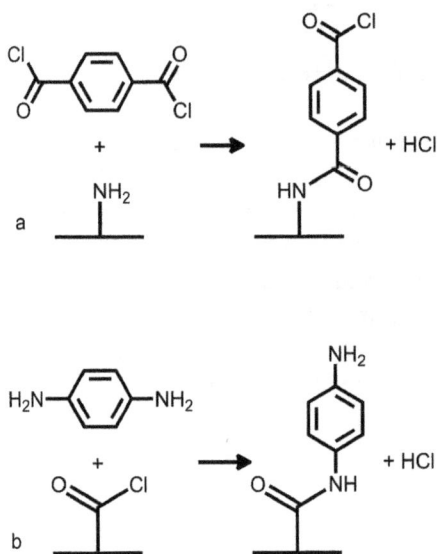

Figure 8.9: Polymerization of EDOT through single electron oxidation.

The growth mechanism is oxidative, where the second precursor ($MoCl_5$) oxidizes the EDOT, which then allows cross-linking between two oxidized EDOT moieties. Notably, these two linked monomers can still each be oxidized in a subsequent step and lose their second hydrogen atom to crosslinking. This is the ongoing long chain formation mechanism to make PEDOT.

This is an interesting process to have developed into an MLD process: there is no obvious linear chain growth, but rather the first pulse would deliver a monolayer of the EDOT monomer to the surface, and the second pulse causes the oxidation, and this is where crosslinking and polymer chain formation occurs. In the growing film, it is easy to imagine that EDOT monomers can adsorb to the growing film through hydrogen bonding (Figure 8.10), and then (once oxidized), these can crosslink to the layer below.

Again, this weak, hydrogen-bonding nucleation of EDOT is a physisorbed state, and (as expected), the growth per cycle decreases noticeably with increased temperature. The single electron oxidation step by $MoCl_5$ is likewise difficult to imagine as an MLD half-cycle, but easier to consider as the redox half-reactions:

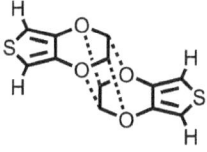

Figure 8.10: Potential adsorption of EDOT to a growing film.

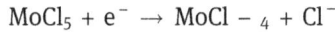

$$MoCl_5 + e^- \rightarrow MoCl_{-4} + Cl^-$$

$$EDOT \rightarrow EDOT^+ + e^-$$

This is an elegant selection of oxidizing agent, as both molybdenum chloride species are volatile and able to participate as a precursor and by-product of the oxidizing pulse, respectively. Notably, the PEDOT will contain chloride anions, which cannot volatilize away. This is an inherent impurity issue for this MLD process.

Chapter 9
ALE processes

Etching is an important step in pattern transfer processes, used in micro- and nanoelectronics. The main etch process that is typically used is a reactive ion etch, where ion sources remove material in a fast, anisotropic (i.e., directional), and continuous manner. Naturally, the desire to evolve etch as a "reverse" ALD process was desirable: to etch a surface with Angstrom precision in a layer-by-layer manner (Figure 9.1).

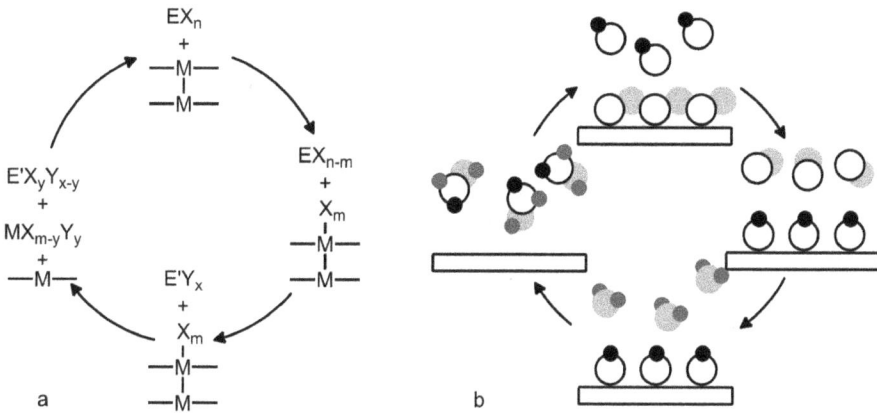

Figure 9.1: The depiction of an ALE cycle to mirror the ALD cycle found in Chapter 1, where a) shows a stoichiometric balanced cycle, and b) shows a cartoon depiction.

ALE requires a step where the element or elements available at the surface are chemically activated by chemisorption, whether by ligands or by a precursor. The second step of an etch process is to enable the activated surface to form volatile by-products that include the surface elements of the film. In this chapter, thermal ALE will be examined, and so this step in Figure 9.1a is through a ligand exchange between the surface and a precursor. The stoichiometry is again difficult to take in on first reading, but it represents the idea that the surface needs a minimum of one ligand or group from each step ultimately form a gas-phase by-product. Likewise, the cartoon depiction is not a familiar as it is for the case of ALD. Figure 9.1b depicts an etch where each precursor pulse removes a component of the target film, and the preserved

Note: In this chapter, I would like to again acknowledge Steven George, who has been a main motivator of this field of research. Also, Thorsten Lill and the Lam Research Corporation deserve recognition here. This company has done a significant amount of development in atomic layer etch, and their work has produced an excellent review article on thermal ALE.

https://doi.org/10.1515/9783110712537-009

stoichiometry of the "circles" demands some imagination when it comes to the etch chemistry. However, both depictions are worth considering to compare the surface chemistries of ALD and ALD. Notably, ALE can be quite variegated, and each pulse in an ALE process does not necessarily have to remove an element of the target film; however, for etch to be complete, all film components must be removed at the end of one ALE cycle.

Notably, thermal ALE is isotropic (i.e., non-directional), like ALD. The mechanism allows for the precursors to reach all surfaces and (ideally) etch them equally. This can be differentiated from ALE having an ion etching step, where the ion beam can etch only in the "line of sight" of the beam, and therefore undergoes directional etching. Thus, thermal ALE cannot replace all etching steps in microelectronics, and ALE with anisotropic steps as part of the process is also necessary to have a complete toolbox of atomic layer processing techniques. But for the purpose of examining surface chemistry, thermal ALE offers several interesting examples.

Atomic layer etch (ALE) cannot simply occur by running ALD processes in reverse: the fact that thermodynamic spontaneity drives ALD to completion means that the direct reverse of an ALD process would be thermodynamically prohibited. Rather, the surface chemistry of ALE processes has to be carefully designed to be thermodynamically spontaneous. One of the first ALE processes used the chemisorption of Cl_2 gas at a gallium arsenide surface to create a chemically activated surface layer:

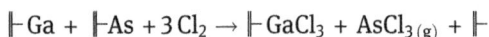

$$\vdash Ga + \vdash As + 3\,Cl_2 \rightarrow \vdash GaCl_3 + AsCl_{3\,(g)} + \vdash$$

The trichloroarsenic(III) was found to desorb quickly and spontaneously from the surface, but energy was needed to desorb the trichlorogallium(III) into the gas phase, and so laser ablation was used. Importantly, this demonstrates that the role of an effective ALE precursor is to form a new surface on the target film that is chemically more reactive and that permits a volatile species to form.

9.1 Etching oxides

Aluminum oxide can be etched by the addition of HF and trimethylaluminum(III) (Figure 9.2).

In the first step, HF etches surface hydroxyl and oxo groups to produce water as a by-product and form a fluorine-terminated aluminum surface (Figure 9.2a). The HF likely coordinates to a surface oxo bridge through the proton, forming a hydroxyl group and opening the bridge. This creates an electrophilic aluminum defect that then immediately bonds to the fluoride. Notably, the surface does not need hydroxyl groups to initiate this reaction, but after a hydroxyl is formed, a second HF

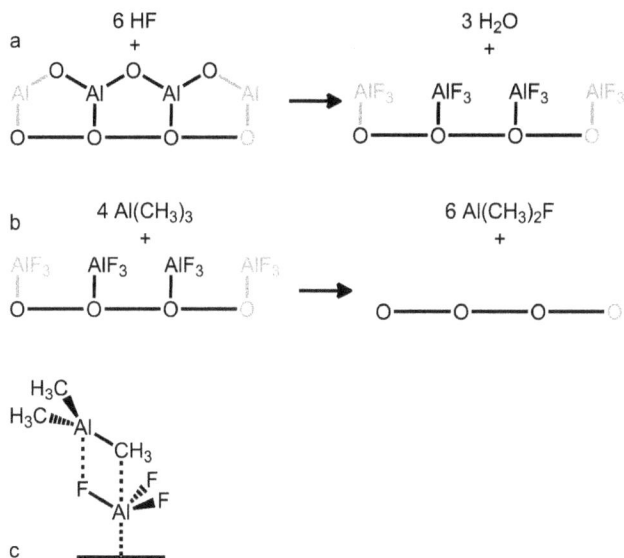

Figure 9.2: Etching alumina using HF and TMA.

molecule is needed to free the oxygen from the surface. Thus, this step of the etch is effective on both hydroxylated and non-hydroxylated alumina surfaces.

In this instance, the formation of trifluoroaluminum(III) is not spontaneous, and so self-limiting saturation occurs, and the thermodynamically stable by-product water is produced. The second step is a ligand exchange reaction with physisorbed trimethylaluminum(III), forming dichlorofluoroaluminum(III), which is volatile and released from the surface (Figure 9.2b). This etches the aluminum from the surface, reforming a surface with available oxo groups for the cycle to continue. The ligand exchange at the surface is likely a one-step process since the two aluminum centers would have little difference in electronegativity, leading to neither of them preferentially undergoing dissociation to free a ligand (Figure 9.2c). This is an elegant example of etching, in that the binary surface (i.e., containing both oxygen and aluminum) is etched through loss of one component of the surface with each reaction. This ALE process, where the surface is fluorinated and water is evolved, followed by ligand exchange with TMA, can be generalized over several different oxides (VO_2, ZnO, Ga_2O_3, ZrO_2, HfO_2). This etch can also be exploited for other oxides, by first converting the surface to aluminum oxide, followed by etching. For example, silicon oxide can be etched by first making a more chemically reactive surface through reaction with TMA:

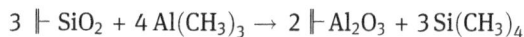

$$3 \vdash SiO_2 + 4\,Al(CH_3)_3 \rightarrow 2 \vdash Al_2O_3 + 3\,Si(CH_3)_4$$

Once Al_2O_3 is formed, the sequence of HF followed by TMA can etch this layer, exposing the SiO_2 underneath.

9.2 Etching metals

One common strategy for etching metals is to convert the surface into a metal oxide, and then etch the oxide. A good example of this is the ALE of tungsten metal:

$$\vdash W + O_3 \rightarrow \vdash WO_3$$

The oxidant can be either dioxygen or ozone gas, and this is not truly a self-limiting reaction. Rather, the longer the exposure, the deeper the tungsten is oxidized, until the diffusion of the oxygen source is kinetically limited from further penetration past the surface.

What is self-limiting however, is the etching of the WO_3 layer with trichloroboron(III) (Figure 9.3).

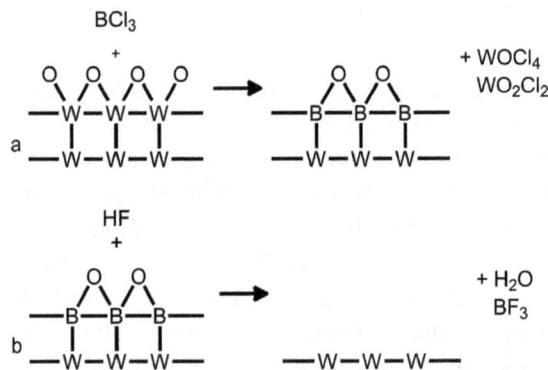

Figure 9.3: The thermal ALE of WO_3 using trichloroboron(III) and HF.

The oxide layer can then be exposed to trichloroboron(III) to form a mixture of volatile tungsten oxychloride species and a film of B_2O_3 (Figure 9.3a). This etch step does not follow a single dominant reaction path, but rather produces a family of tungsten oxychlorides of the general formula WO_xCl_{6-2x}. Here, the tungsten maintains an oxidation state of + 6, and each W^{6+} accumulates oxygen and chloride to form a neutral compound, and these then volatilize from the surface.

Subsequently, the B_2O_3 is etched with HF, forming H_2O and BF_3 and revealing the remaining WO_3 that was not converted on trichloroboron(III) exposure (Figure 9.3b). In every cycle, it is then necessary to reoxidize the surface to form the WO_3. Although the conversion of the surface with trichloroboron(III) and subsequent etch with HF are both self-limiting step, the inherent lack of self-limiting behavior of the tungsten oxidation means that there can be a variable thickness of WO_3 remaining. Without the (non-self-limiting) oxidation step, this etch chemistry cannot go forward reliably.

Interestingly, the oxide formation is kinetically self-limited when oxygen is used. In this case, the trichloroboron(III) isn't necessary, and tungsten oxide can be directly etched with hexachlorotungsten(VI) to again form WO_xCl_{6-2x} species:

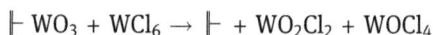

$$\vdash WO_3 + WCl_6 \rightarrow \vdash + WO_2Cl_2 + WOCl_4$$

This method can also be used to etch silicon and silicon nitride (although these are not metals). The formation of a thin, self-limiting oxide passivation layer is the key to this chemistry. The formation of passivating SiO_2 at Si surfaces can reach anywhere between 5 and 10 Å at room temperature in a week, depending on the doping of the Si^0 as well as environmental factors like relative humidity. This can be drastically accelerated by heating and through using alternate oxidation precursors like O_3 or oxygen plasma:

$$\vdash Si + 2O\cdot \rightarrow \vdash SiO_2$$

This layer can be converted into diammonium hexafluorosilicate by exposure to a plasma of NH_3/NF_3:

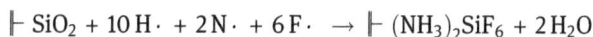

$$\vdash SiO_2 + 10 H\cdot + 2 N\cdot + 6 F\cdot \rightarrow \vdash (NH_3)_2SiF_6 + 2 H_2O$$

Although written stoichiometrically here, it should be obvious that this "reaction" is due to a variety of complicated surface reactivity that settles into the thermodynamically stable formation of $(NH_3)_2SiF_6$. This salt can be thermally decomposed at 150 °C. It thermally decomposes to HF and SiF_4, as well as silicon-nitrogen containing species. Thus, the etch step needs to be activated briefly by heat in order to undergo the second step in this ALE process. If the entire process were run at temperatures where the decomposition of $(NH_3)_2SiF_6$ underwent spontaneous thermal decomposition (i.e., > 150 °C), the process would undergo continual "CVD-like" etch.

Metal surfaces can also be etched directly using dihalogen gases. In this case, two different ligands are necessary, as shown in the general ALE cycle in Figure 9.1. The initial reaction requires chemisorption of the dihalogen at the metal surface to form a metal halide surface. The second ALE step in this case is the introduction of the conjugate acid of a ligand to react at the surface to produce a volatile species. A good example of this is type of ALE is the etch of iron by dichlorine gas and acetylacetone (Figure 9.4).

The first step is straightforward, where the metal surface is chlorinated by the Cl_2, which is formally an oxidative addition of Cl_2 to the surface (Figure 9.4a). The introduction of Hacac (the conjugate acid of the acetylacetonate ligand) in the second step of this ALE process is more complicated. The iron species formed through reaction with

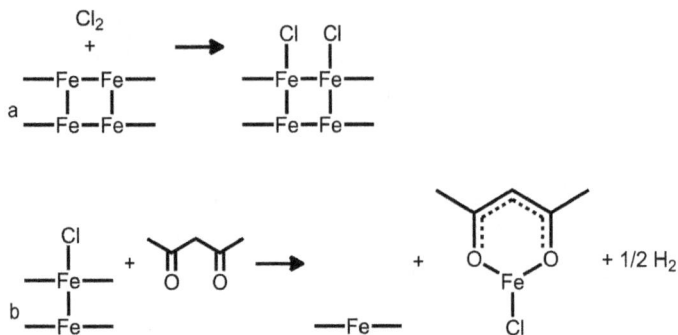

Figure 9.4: Etching iron with dichlorine and acetylacetone.

Hacac could have a Fe^{2+} or Fe^{3+} oxidation state. Given that there are two ligands (chloride and acac), and two potential oxidation states, many different iron-containing volatile by-products could be formed. This lack of a direct reaction pathway is very common in surface mechanism studies, and it can be difficult to guess the nature of the surface chemistry without additional data. In this case, chemical modeling suggests that a heteroleptic Fe(II) compound is the most likely by-product (Figure 9.5).

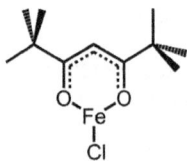

Figure 9.5: Acetylacetonatochloroiron(II) is the likely etch by-product of etching iron metal with Cl_2 and Hacac.

The proton from Hacac is likely left behind at the surface, and will react either with remaining, surface bonded chlorides to produce volatile HCl, or it will undergo reductive elimination with another surface hydride to form H_2.

This same etch method can be used to etch copper and cobalt metal. In these cases, the metal surface is activated by reaction with O_2, and the oxide subsequently etched by the conjugate acid hexafluoroacetylacetone (Hhfac, $F_3C(C=O)CH_2(C=O)CF_3$):

$$\vdash MO + 2\,Hhfac \rightarrow \vdash + H_2O + M(hfac)_2, \ M = Co^{2+}, Cu^{2+}$$

The step with Hhfac etches both the metal away in the neutral compound $M(hfac)_2$, and the oxygen away as water. This is a straightforward example of metal ALE, and it requires the oxidation of the metal surface to be self-limiting. Like the example of tungsten etching presented earlier, if the oxide layer is formed deeper than one monolayer without a kinetic self-limitation, the ligand step will continue to etch if there is available metal oxide. This is a pitfall of employing surface oxidation as the step for activating a metal surface to etch.

The field of ALE is still growing, and the examples here show some strategies to exploit the specific surface chemistry, and metal-ligand chemistry to etch a variety of films.

Index

https://doi.org/10.1515/9783110712537-010